Intelligent Mobile Malware Detection

The popularity of Android mobile phones has caused more cybercriminals to create malware applications that carry out various malicious activities. The attacks, which escalated after the COVID-19 pandemic, proved there is great importance in protecting Android mobile devices from malware attacks. **Intelligent Mobile Malware Detection** will teach users how to develop intelligent Android malware detection mechanisms by using various graphs and stochastic models. The book begins with an introduction to the Android operating system accompanied by the limitations of the state-of-the-art static malware detection mechanisms as well as a detailed presentation of a hybrid malware detection mechanism. The text then presents four different system call-based dynamic Android malware detection mechanisms using graph centrality measures, graph signal processing and graph convolutional networks. Further, the text shows how most of Android malware can be detected by checking the presence of a unique subsequence of system calls in its system call sequence. All the malware detection mechanisms presented in the book are based on the authors' recent research. The experiments are conducted with the latest Android malware samples, and the malware samples are collected from public repositories. The source codes are also provided for easy implementation of the mechanisms. This book will be highly useful to Android malware researchers, developers, students and cyber security professionals to explore and build defense mechanisms against the ever-evolving Android malware.

Security, Privacy, and Trust in Mobile Communications

Series Editor: Brij B. Gupta

This series will present emerging aspects of the mobile communication landscape, and focuses on the security, privacy, and trust issues in mobile communication-based applications. It brings state-of-the-art subject matter for dealing with the issues associated with mobile and wireless networks. This series is targeted for researchers, students, academicians, and business professions in the field.

Nanoelectric Devices for Hardware and Software Security

Balwinder Raj & Arun Kumar Singh

Cross-Site Scripting Attacks: Classification, Attack, and Countermeasures

B. B. Gupta & Pooja Chaudhary

Smart Card Security: Applications, Attacks, and Countermeasures

B. B. Gupta & Megha Quamara

Computer and Cyber Security: Principles, Algorithm, Applications, and Perspectives

Brij B. Gupta

Intelligent Mobile Malware Detection

Tony Thomas, Roopak Surendran, Teenu S. John & Mamoun Alazab

Intelligent Mobile Malware Detection

Tony Thomas, Roopak Surendran
and Teenu S. John
Kerala University of Digital Sciences, Innovation and Technology
(Digital University of Kerala), Kerala, India

Mamoun Alazab
College of Engineering, IT and Environment at
Charles Darwin University, Australia

CRC Press
Taylor & Francis Group
Boca Raton London New York

CRC Press is an imprint of the
Taylor & Francis Group, an **informa** business

First edition published 2023
by CRC Press
6000 Broken Sound Parkway NW, Suite 300, Boca Raton, FL 33487-2742

and by CRC Press
4 Park Square, Milton Park, Abingdon, Oxon, OX14 4RN

CRC Press is an imprint of Taylor & Francis Group, LLC

© 2023 Tony Thomas, Roopak Surendran, Teenu S. John, and Mamoun Alazab

ISBN: 978-0-367-63871-9 (hbk)
ISBN: 978-1-032-42109-4 (pbk)
ISBN: 978-1-003-12151-0 (ebk)

DOI: 10.1201/9781003121510

Typeset in Erewhon font
by KnowledgeWorks Global Ltd.

Contents

Preface

Nowadays, smart phones are widely used for making phone calls, sending messages, storing personal data, browsing the Internet, online banking and more. Because of this, smart phones have become targets for cyber-attacks involving malware. Cyber criminals are targeting smart phones to spread malware in order to steal money and confidential data stored in those phones. Malware applications such as trojan SMS and trojan banker can cause great financial loss to users. Trojan SMS can send SMS messages to premium rate numbers in the background and trojan banker can steal the online banking details of a user without the user's knowledge. Therefore, it has become very essential to secure smart phones against malware attacks.

With the widespread usage of the Android operating system, the number of malware targeting Android smart phones has risen several folds. Almost 98% of smartphone malware are designed for Android devices. Most of the existing anti-malware products are still relying on static and signature-based malware detection mechanisms. Static analysis is a method of detecting malware application by analyzing the source code of the application without executing it. In signature-based analysis, the hash value of an application is compared with a list of hash values of known malicious applications for identifying whether the application is one among the listed malware. These detection mechanisms can be easily evaded by code transformation attacks. Hence, it is essential to develop novel malware detection mechanisms based on dynamic analysis for accurate malware detection. Dynamic analysis mechanisms consider runtime information such as system metrics, network level information, system calls and more for detecting the malicious behavior of the application. A malicious application typically invokes sensitive APIs in an automated manner to perform privileged operations. This automated invocation of API calls gets reflected in the system call sequence of the application. Hence, system calls are considered as one of most effective features for capturing the malicious behavior of an application.

Most of the existing system call-based malware detection mechanisms consider the system call frequencies or co-occurrences in the system call sequence for malware detection. The system call frequency-based mechanisms use machine learning classifiers to detect malware based on the independent occurrences of each individual system call in the entire sequence. These mechanisms do not consider the relationships among the system calls in a system call sequence. In system call co-occurrence-based mechanisms, the mutual relationships between system calls in the sequence are considered for malware detection. However, these approaches do not consider the complex relationships among the system calls crucial for identifying the malicious behavior of an application.

This book is an attempt to present representation and characterization of Android malware using graph and stochastic models and use such representations and characterizations to detect Android malware. First, the state-of-the-art static malware detection mechanisms and their limitations are presented in this book. This will be followed by detailed presentations of a hybrid malware detection mechanism and four different system call-based dynamic Android malware detection mechanisms based on recent research by the authors. This book will teach readers how to develop effective Android malware detection mechanisms using graph centrality measures, graph signal processing and graph convolutional networks. The source codes are also provided in the appendix for easy implementations of the mechanisms. This book will be highly useful for Android malware researchers, developers, students and cyber security professionals.

In Chapters 1 and 2, the basics of Android OS and Android malware are discussed. In Chapter 3, state-of-the-art static malware detection mechanisms and their limitations are presented. In Chapter 4, a tree augmented naive (TAN) Bayes-based hybrid malware detection mechanism, which uses the conditional dependencies among relevant static and dynamic features (API calls, permissions and system calls) required for the functionality of an application, is presented. Three ridge regularized logistic regression classifiers corresponding to API calls, permissions and system calls of an application are used along with the TAN model for identifying whether the application is malicious or not. In Chapter 5, a malware detection mechanism, which uses machine learning classifiers on various centrality measures calculated from the system call digraph of an application, is presented. In Chapter 6, the graph convolutional neural (GCN) network is used to detect the malicious behavior from the system call digraphs. In Chapter 7, a way to construct low-dimensional feature vectors (graph signals) from system calls using graph signal processing (GSP) is described. These graph signals are used as feature vectors of machine learning classifiers for identifying the malicious behavior. Through the implementations of these methods, it is shown that graph-based mechanisms are very accurate and efficient in detecting malware applications over traditional mechanisms. In Chapters 4–7, machine learning classifiers are used to detect malware. The main problem of a machine learning approach is the difficulty in finding the properties or features that uniquely characterize the Android malware. Toward this, in Chapter 8 it is shown that most of the Android malware could be detected by checking the presence of a unique short system call subsequence (malicious system call code) in its system call sequence. This detection mechanism does not require any machine learning classifiers. Through experiments, the existence of malicious system call code is shown in the majority of malware applications that use the system resources in the background. The book concludes with Chapter 9, which includes conclusions, limitations and future directions for research.

Acknowledgements

We thank the Government of Kerala for supporting this project through the Kerala State Planning Board project CRICTR.

We owe our sincere gratitude to our colleagues at the Kerala University of Digital Sciences, Innovation and Technology, Indian Institute of Information Technology and Management-Kerala, and Charles Darwin University, who gave us the time and space to complete this book.

This book benefited from the insights and direction of several people including the anonymous reviewers of our research papers. We use this opportunity to express our heartfelt thanks to all who supported and indirectly contributed to this book project.

We thank the publishers, CRC Press/Taylor & Francis, for the valuable support provided throughout the entire process.

We thank our family and friends for their support in helping to make this book possible.

Finally, we thank all our readers for their interest in Android malware analysis and investing their time in reading this book.

About the Authors

Dr. Tony Thomas is currently associate professor in the School of Computer Science and Engineering, Kerala University of Digital Sciences, Innovation and Technology, India (formerly IIITM-K). He completed his master's and PhD degrees from IIT Kanpur. After completing his PhD, he carried out his post-doctoral research at the Korea Advanced Institute of Science and Technology. After that, he joined as a researcher at the General Motors Research Lab, Bangalore, India. He later moved to the School of Computer Engineering, Nanyang Technological University, Singapore as a research fellow. In 2011, he joined as an assistant professor at Indian Institute of Information Technology and Management-Kerala (IIITM-K). He is an associate editor and reviewer of several journals. He is a member of the Board of Studies of several universities. His current research interests include: malware analysis, biometrics, cryptography, quantum computation and machine learning applications in cyber security. He has published many research papers, book chapters and books in these domains. He is an author of the book *Machine Learning Approaches in Cyber Security Analytics* published by Springer.

Dr. Roopak Surendran is currently working as a penetration tester at the Kerala Security Audit and Assurance Centre (K-SAAC) of the Kerala University of Digital Sciences Innovation and Technology. He has done his PhD research in Android malware analysis, which was funded by the Kerala state planning board. Before joining the PhD program, he completed his MPhil degree in computer science with a specialization in cyber security from Indian Institute of Information Technology and Management-Kerala. He published many research papers related to malware analysis and phishing detection. Also, he has developed Python-based tools and sandboxes to protect devices from phishing and malware attacks. His interests include: web application security, mobile application security, malware analysis and phishing detection.

Ms. Teenu S. John holds an MTech degree in computer science with specialization in data security from TocH Institute of Science and Technology under Cochin University of Science and Technology, Kerala, India, and a BTech degree in information technology from the College of Engineering Perumon, under Cochin University of Science and Technology-Kerala, India. She is currently doing her PhD on adversarial malware detection at the Kerala University of Digital Sciences Innovation and Technology, formerly Indian Institute of Information Technology and Management-Kerala (IIITM-K). Her research interests include: malware analysis, machine learning for cyber security, data analytics and cyber threat detection.

Dr. Mamoun Alazab is associate professor at the College of Engineering, IT and Environment, and is the director of the NT Academic Centre for Cyber Security and Innovation (ACCI) at Charles Darwin University, Australia. He received his PhD in computer science from the Federation University of Australia, School of Science, Information Technology and Engineering. He is a cyber security researcher and practitioner with industry and academic experience. Dr. Alazab's research is multidisciplinary focusing on cyber security including current and emerging issues in the cyber environment like cyber-physical systems and Internet of Things, with a focus on cybercrime detection and prevention. He has more than 300 research papers, 11 authored and edited books, as well as 3 patents. As of March 2022, 9256 citations appear on Google. His research over the years has contributed to the development of several successful secure commercial systems. His book, *Malware Analysis Using Artificial Intelligence and Deep Learning*, reached 40k downloads in about 1 year and was referred to by Microsoft research and Google research. He is the recipient of several prestigious awards including the NT Young Tall Poppy of the Year (2021) from the Australian Institute of Policy and Science (AIPS) and the Japan Society for the Promotion of Science (JSPS) fellowship through the Australian Academy of Science. He worked previously as a senior lecturer (Australian National University) and lecturer (Macquarie University). He is a senior member of the IEEE, and the founding chair of the IEEE Northern Territory (NT) Subsection. He serves as the associate editor of *IEEE Transactions on Computational Social Systems, IEEE Transactions on Network and Service Management (TNSM), ACM Digital Threats: Research and Practice*, and *Complex & Intelligent Systems*.

Symbols

α	Attenuation factor	$C_1(.)$	Eigen vector centrality
β	Damping factor	$C_2(.)$	Betweenness centrality
γ	Penalty value	$C_3(.)$	Closeness centrality
δ	Regression parameter	$P(.)$	Page rank
λ	Largest eigen value of the adjacency matrix	S	Set of system calls
		S_i	i^{th} system calls
ϵ	Information distance value	$\mathcal{Y}$	System call sequence
μ	Stationary distribution	X	Refined system call sequence
$Pr(.)$	Probability	G	System call graph
$\mathcal{H}(.)$	Entropy	E	Edges of system call graph
$\mathcal{A}$	Permission-based feature vector	$\mathcal{A}$	Adjacency matrix of system call graph
$\mathcal{B}$	API call-based feature vector	$\mathcal{N}$	Neighbourhood matrix of system call graph
C	System call frequency-based feature vector		
$\mathcal{D}$	Labeled dataset	T	Transition probability matrix
$\mathcal{E}$	Label of an application	Q	Rank 1 matrix from T
$\mathcal{L}_1$	API call classifier	V_0	Graph signal vector
$\mathcal{L}_2$	Permission classifier	V_1	Transformed graph signal vector
$\mathcal{L}_3$	System call classifier		
$\mathcal{T}_1$	Threshold for $\mathcal{L}_1$	$\tilde{h}_{S_i}^t$	Hidden vector of S_i
$\mathcal{T}_2$	Threshold for $\mathcal{L}_2$	W^t	Weight matrix
$\mathcal{T}_3$	Threshold for $\mathcal{L}_3$	z_{S_i}	Node embedding vector of S_i
$\mathcal{T}$	Threshold for TAN model	Tr	Training set of graphs
$\mathcal{Z}_1$	Output of $\mathcal{L}_1$	lb_i	Label of ith graph
$\mathcal{Z}_2$	Output of $\mathcal{L}_2$	σ_{ji}	Length of shortest path from j^{th} vertex to i^{th} vertex
$\mathcal{Z}_3$	Output of $\mathcal{L}_3$		
$K(.)$	Katz centrality	$\mathcal{P}$	System call patterns

1

Internet and Android OS

Android operating system is the most widely used mobile operating system dominating the global market share. In this chapter, we provide an introduction to Android operating system and its architecture. We also discuss about why Android operating system is widely adopted for developing smart and scalable Internet of Things (IoT) infrastructure. Towards the end of the chapter, we also provide a brief discussion about the security of IoT and also about the various malware attacks in IoT till date.

Android operating system (Android OS) was developed in Palo Alto of California in 2003 [136]. It is based on the Linux kernel and it is available as an open source software. Due to the open source nature, the developers find it easy to build new applications with varied functionalities. Android OS has dominated the mobile operating system market with a share of 74.14% in June 2020 which clearly shows its widespread usage. Figure 1.1 shows the market share of mobile operating systems worldwide. The figure clearly shows the popularity of Android OS when compared to other mobile operating systems.

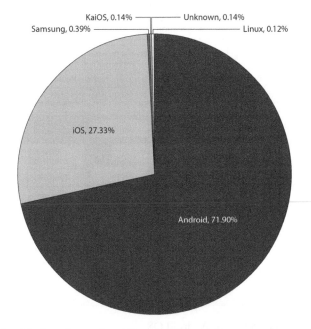

FIGURE 1.1 Market share of mobile operating systems.

DOI: 10.1201/9781003121510-1

1

The first version of Android known as Android 1.0 was released in September 2008. Since then, several versions have been released with various features[170]. The following section discusses about the architecture of Android OS in detail.

1.1   Android OS

Android OS comprises of a stack of software components. Figure 1.2 shows its architecture. There are five layers in Android OS. They are:

1. Linux Kernel;

2. Libraries;

3. Android Runtime;

4. Application Framework; and

5. Applications.

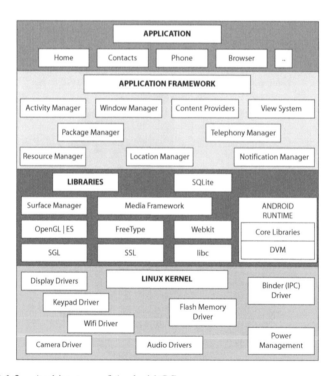

FIGURE 1.2   Architecture of Android OS.

Now, we will discuss about these layers in detail in the following sections.

1.1.1   Linux kernel

In Android, the Linux kernel constitutes the bottom layer of the software stack. The Linux kernel can support hardware drivers and it is used to manage input and output requests from the user. The device drivers present in the linux kernel are software used to communicate to a particular device. For example, if we need to access the camera and take the photos, the driver will give necessary commands to the camera hardware to do so. The linux kernel also provides wide range of functionalities such as process management, memory management, device management, etc. Each version of the Android has its own linux kernel version. Android uses the kernel version 4.4 or 4.9 or 4.14 or 5.4 or 5.10 as of 2022.

1.1.2   Native libraries

On top of the linux kernel, there exists the native libraries layer which consists of various C/C++ core libraries and java-based libraries such as SQLite, WebkitSurface Manger, OpenGL, SSL, libc, Graphics, Media, etc. These libraries are used for handling different types of data. The surface manager library is used to manage the display while the SGL and openGL libraries are used to manage 3D and 2D graphics. The media library is used for managing and storing video and audio files and SQL library is used for database support and WebKit library is used for web browser support. The SSL library is used to manage the security. The FreeType library is used for font support while libc library is used to support C libraries.

1.1.3   Android runtime

Android runtime (ART) consists of core libraries and Dalvik Virtual Machine (DVM). ART forms the basis of the application framework and it helps to launch an Android application with the libraries. The DVM is a java based virtual machine and it helps the Android application to run its own instances of the virtual machine. The DVM also helps to run multiple instances of the virtual machine simultaneously assuring memory management, security, isolation and threading. The DVM also helps to run the .dex files created from .class files. The core libraries present in the ART helps to create Android applications using the Java language.

1.1.4   Application framework

Application Framework provides services to the application in the form of Application Programming Interfaces (API's). The Application Framework consists of the following components:

- Activity Manager: The Activity Manager is used to manage the life cycle of applications.

- Content Providers: The Content Providers manage the data sharing between applications.

- Telephony Manager: The Telephony Manager manages all the telephony related functionalities.

- Location Manager: The Location Manager is used to get periodic updates about the device's geographical location.

- Resource Manager: The Resource Manager manages the various types of resources used in application.

1.1.5 Application layer

The application layer constitutes the topmost layer. This layer consists of Android applications which are prebuilt in the system like SMS client app, Contact manager and also the customized applications build by the developer. The applications make use of the services of the application framework layer to build the required applications. Android applications are developed in Java programming language[173][148]. The main components of an Android application are given below:

- Activity: Activities provide a user interface to interact with the Android application.

- Services: This component deals with the background operations such as uploading/downloading the data. The service component does not have a user interface.

- Broadcast receiver: This component deals with the external events such as incoming SMS, reboot the device, etc.

- Content provider: It provides a consistent interface for data access between different apps.

1.2 Android Application Development

The code of the Android applications contains the necessary components required for the functionality of the application. Initially, the Java source code is converted into the byte code (.class file) using the Java compiler. Then, the byte code (.class file) is converted to dalvik executable format (.dex file) using dx tool [7]. Finally, aapt tool [6] is used for converting .dex file into Android application (.apk file).

The development stages of an Android application is given in Figure 1.3.

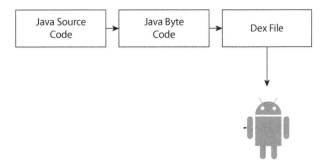

FIGURE 1.3 Android application development.

1.3 Google Playstore

Google play store is the official repository of Android applications. Developers can submit applications to Google playstore for publishing them. The submitted applications undergo security checks via a built-in mechanism called bouncer [159][198][145]. The algorithm of bouncer is not revealed by Google. Bouncer executes the submitted application in a virtual machine and checks for the malicious activities. Google play publishes an application only if no malicious activity is detected by the bouncer. After publication, a user can directly install the application into his/her device. There are several categories of Android applications found in Google play. The sample categories of apps in Google play are given below.

- Gaming apps

- Sports related apps

- Social networking apps

- Banking apps

- Education apps

- Communication apps

- Photography apps

- News and magazine apps

- Weather apps

- Parenting apps etc.

1.4   Intents and Intent Filters

Intents are objects used to interact with the components of the same applications and also with the components of other applications. There are two types of intents, namely explicit intents and implicit intents [32]. They are described below.

- Explicit intents: Explicit intents are typically used to start a component in the particular application itself. For example, to start a new activity within the application in response to a user action.

- Implicit intents: Implicit intents declare a general action to perform and it helps a component from another application to handle it. For example, if an application wants to show the user a location on a map, then we can use an implicit intent that requests some other application like Google Maps to show the location.

An application can permit other applications to access its components (activity, service, broadcast receiver, etc) using intent filters. The elements of the intent filters are:

- Action: It indicates the type of action performed by the invoking component. For example, ACTION_VIEW is used for viewing the contents.

- Data: It indicates the type of data received by the intent.

- Category: It specifies the launching location of the component.

1.5   Android Security

In Android, a privilage escalation model is implemented to ensure that an application cannot access other application's code or data. Android application security is achieved using permissions, application sandbox, application signature and data encryption [26].

1.5.1   Permissions

In order to restrict an application from accessing the sensitive functionalities of a device such as telephony, network, contacts, sdcard and location, Android provides a permission-based security model in its application framework[36]. Android permissions are classified into four protection-levels. They are normal, dangerous, signature and signatureOrSystem. Normal permissions are granted default during the installation time of the application. Dangerous permissions are considered as high risk permissions since they have the capability to access the private data and important device sensors such as camera. Signature permissions are the permissions which are granted only if the requesting application is signed with the same developer certificate. They are granted automatically at the installation time. SignatureOrSystem permissions are granted only if the application is signed with the same certificate as the Android system image. They are also granted automatically during the installation time.

1.5.2 Application sandbox

Android uses the Linux operating system based protection to ensure that the resources of each application is isolated. It also provides unique User Identification (UID) to ensure isolation. Each application is also made to run in its own DVM or Android Runtime (ART). The processes communicate with each other with the help of Binders. Binders act as an interprocess communication systems as well as remote method invocation systems. To communicate with other processes, the application sends messages to the Binder that checks the Activity Manager to verify whether the application has permissions to communicate.

1.5.3 Application signature

Android application signature ensures that the application has not been manipulated by the malware developers. If an application is decompiled, the signature is no longer considered as valid. The signing is done by the application developers using a certificate. Android provides three schemes for generating application signature. They are given below[14].

- APK Signature Scheme v1: The v1 scheme is based on JAR signing. However v1 signing scheme can only protect some parts of the application.

- APK Signature Scheme v2: The v2 signature scheme was introduced in Android 7.0. It has faster application installation and it also offers more protection to unauthorized apk manipulations.

- APK Signature Scheme v3: The v3 signature scheme was introduced in Android 9. The format of v3 signature scheme is similar to that of v2. However, the v3 signature scheme adds information about the supported SDK versions.

1.5.4 Data encryption

Android also ensures data encryption to protect users data [21]. Android uses symmetric encryption to secure the user's data. In Android, there are two types of encryptions. They are file-based encryption and full disk encryption. File-based encryption is supported by Android 7.0 and later. In this encryption, each file is encrypted with a different key. Android 9 uses metadata encryption with hardware support. In metadata encryption, the contents that are not encrypted by file based encryption such as directory layouts, file sizes, permissions, etc. are encrypted with a single key. This key is protected by the KeyMaster that is again protected by Verified Boot. The Verified Boot ensures all executed code comes from a trusted source (usually device OEMs), rather than from an attacker or corruption. In full disk encryption, a single key is used to encrypt all the files and the key is protected by the device password. This type of encryption is supported by Android 5.0 to Android 9.

1.6 Internet of Things

The Internet of things (IoT) describes the network of physical objects or things that are embedded with sensors, software, and other technologies for the purpose of connecting and exchanging data with other devices and systems over the Internet[33]. With IoT, billions of objects can sense, share the information and can take decision of their own. This section discusses about the various components of IoT[163] and also why Android Things help to build scalable and secure IoT infrastructures.

1.6.1 Architecture of IoT

The architecture of IoT comprises of various technologies that support each other to achieve scalability, availability, maintainability and functionality. Android architecture comprises of sensors, gateways and networks, management services and applications. Figure 1.4 shows the architecture of IoT. The different layers of IoT are Sensor Layer, Gateways and Networks Layer, Management Service Layer and Application Layer. The details are given in the subsections below.

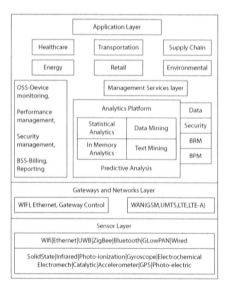

FIGURE 1.4   Architecture of IoT.

1.6.1.1 Sensor layer

The sensor layer is made up of smart objects integrated with sensors that are able to measure temperature, speed, humidity and other factors. In certain cases, these sensors may also possess some memory to record certain measurements. These sensors are then connected to the sensor aggregators in the form of a Local Area Network (LAN) such as Ethernet, WiFi connections or Personal Area Network (PAN) such as ZigBee, Bluetooth and Ultra-Wideband (UWB) [163]. In certain type of sensors, instead of connecting to the sensor aggregates, the connectivity to backend servers/applications can be achieved using Wide Area Network (WAN) such as Global System for Mobile communication (GSM), General Packet Radio Service (GPRS) and Long-Term Evolution (LTE). The sensors with low power and low data rate connectivity forms a Wireless Sensor Network (WSN).

1.6.1.2 Gateways and networks

The data sensed by the sensors are send through the network with the help of gateways and networks. The primary function of this layer is to process the information collected from the sensors and to convert it to digitalized and aggregated versions. To support wider range of applications and transactional services, we can use multiple networks with different access protocols. These networks can be public, private or hybrid which ensures latency, bandwidth and security. The Gateways are implemented with WAN, WiFi or Ethernet.

1.6.1.3 Management service layer

Management service layer constitutes business and process rule engines. The rule engines present in the management service layer helps to take decisions and provide automated processes to obtain more efficient IoT system. This layer also provides Business Rule Management (BRM) and Business Process Management (BPM). It also has various analytical tools to extract information from raw data. Different types of analytics such as in-memory analytics, and streaming analytics are used here. In in-memory analytics, huge volumes of data are cached in the RAM for analysis. Streaming analytics on the other hand, performs the analytics of the data in real-time. This layer also has the ability to control data to reduce the risk of privacy disclosure. In addition to that, various services such as access control, encryption, data management etc. are provided in this layer to ensure security of the data. The Operational Support System (OSS) present in this layer is used to automate the network management function while Business Support System (BSS) supports billing and reporting.

1.6.1.4 Application layer

The application layer consists of various smart applications that work in IoT such as those related to agriculture, factory, supply chain, healthcare, etc.

1.7   Android Things

The main challenges in building an IoT system is interoperability, security and scalability. To solve these issues, Google invented Android Things in 2015. Android Things is an Android-based OS that is meant for IoT. It can address the security issues encountered by the connected devices. The main advantage of Android Things is the easy development environment that makes the developers work on smart displays, digital signboards and Kiosks. Figure 1.5 shows the architecture of Android Things.

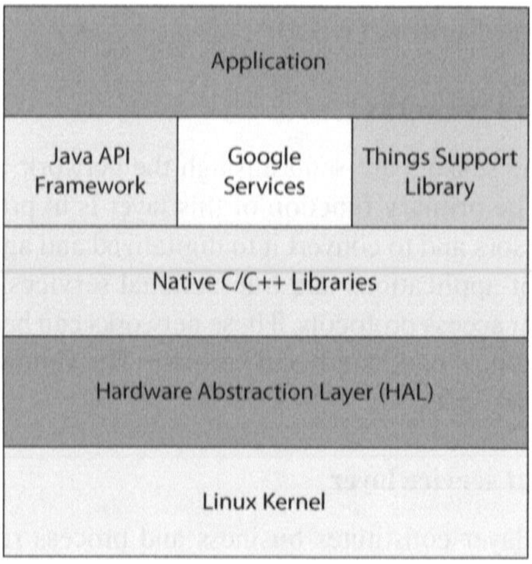

FIGURE 1.5   Architecture of Android Things.

Android Things also solve the problem of interoperability by allowing many devices to work together without any compatability issues. With Android Things, the hardware manufacturers are provided with a certified OS and hardware that makes it easy to build products. Besides that, Android Things also ensures secure IoT infrastructures. This is achieved with Google that sends regular security updates and patches to avoid privacy breaches.

The main difference between Android OS and Android Things is that the system applications and content providers are absent in Android Things. Besides that, Android Things have a Things support library with peripheral I/O API that allows applications to communicate with the sensors and actuators using interfaces and protocols. The Things support library also has a user driver API that allows the applications to inject hardware events. The HAL and the native C/C++ libraries are same as that of the Android operating system. In addition to that, the Google services also offer Google API for Android. The Java API framework provides API's to communicate between the sensors and actuators.

1.8 IoT Security

The popularity of IoT devices has attracted cybercriminals to launch attacks against these systems. In 2018, around 12 million attacks were launched from 69,000 IP addresses. IoT require huge volumes of data for developing smart and intelligent devices. The attacks against IoT can be classified into the three types[34] given below.

- Communication attacks: These attacks target the data transmitted between servers and IoT devices. Attackers may gather the data sent between the IoT devices and servers to access sensitive data.

- Software attacks: Attackers can exploit the vulnerabilities in the web applications and the device software to push malicious firmware updates and steal the credentials.

- Physical attacks: Attackers can target the chip, firmware and physical interfaces to launch their attacks.

1.8.1 Malware Threats in IoT

According to the study of Kasperesky, IoT malware has risen three folds[42]. The most common attack is in the form of botnets. Linux Hydra was the first IoT malware that was capable of launching Distributed denial of service attack. Since then, various malware such as Psybot, Chuck Norris and Tsunami have emerged[53]. Among the various malware attacks against IoT, the Mirai botnet attack affected over 600,000 devices. The attack infected the devices with Argonaut RISC Core (ARC) processors and turned them into a network of bots. Malware such as BrickerBot, VPNFilter, etc. are the variants of Mirai that appeared recently. The common malware infection is through weak Telnet passwords. Table 1.1 shows some of the well known IoT malware families that appeared till the year 2020.

1.9 Conclusion

In this chapter, we discussed about the architecture of Android operating system and also about the IoT architecture in detail. We also discussed about the security threats in IoT and also about the well known IoT malware families appeared till the year 2020.

TABLE 1.1 IoT malware attacks.

Year	Malware	Type
2008	Hydra	Botnet
2009	Psybot/NetworkBluePill	Router Based Botnet
2010	Chuck Norris	Botnet
2011	Umbreon/Umreon/Rebonum/Neobrum	Rootkit
2012	Carna Botnet	Botnet
2012	LightAidra/Linux Aidra	Botnet
2013	Tsunami/Kaiten	Botnet
2013	Linux Darlloz/Zollard	Worm
2014	Gafgyt/BASHLITE/Lizkebab/ Torlus/Qbot/LizardStresser	DDoS Botnet
2014	Spike/Dafloo/MrBlack/Wrkatk/ Sotdas/AES.DDoS	DDoS Botnet
2014	TheMoon	Botnet
2015	Linux Moose/Elan	Worm
2016	VPNfilter	Spyware
2016	Mirai	Botnet
2016	IRCTelnet/LinuxIRCTelnet/NewAidra	Botnet
2017	Amnesia	DDoS Botnet
2017	Linux.MulDrop.14	Trojan
2018	Hide 'N Seek	Botnet
2019	Echobot	Botnet
2020	Mukashi	DDoS Botnet
2020	Rhombus	DDoS Botnet

2

Android Malware

Malware or a malicious software is a program designed to damage or compromise a device or computer system[38].These malicious software can access and damage the valuable cyber assets of the organisation. Eventhough security vendors and researchers are pushing hard to control malware, cybercriminals are finding new ways to evade detection. This chapter explains how the malware has evolved over the past several years. We also explain Android malware types and families and the need for developing secure systems against malware attacks in Android.

2.1 PC Malware vs. Android Malware

The first PC malware, which was virus called *Brain* appeared in 1986[118]. The virus contained hidden copyright messages and it infected the boot sector of the floppy disk. Since then, a variety of malware types and variants have evolved with diverse threat capabilities. The most common ways with which a PC is infected by the malware is through the following mechanisms[151].

1. Spam emails: The most common malware infection is in the form of spam e-mails. The user may receive an e-mail with an attachment stated to contain details of winning a contest, a receipt of an item purchased or an invoice. When the user opens the attachment, it gets downloaded and the malware gets installed into the system.

2. Infected removable drives: This mode of infection is carried out by removable drives such as USB flash drives or external hard drives. These hard drives are installed with potentially dangerous malware that can perform malicious activities on the system.

3. Bundled with other software: In this type of attack, the malware come bundled with legitimate software. These malicious software can be seen in third party websites and peer to peer networks. Some malicious programs disguised as software key generators can also install malware into the PC.

4. Hacked or compromised websites: In this type of attack, when a user access a hacked or compromised website, the vulnerability of the software present in the device can be used by the attacker to infect a malware. Hence

regular software updates are recommended to make the system secure against these threats.

While security vendors are making their products more effective to detect PC malware, cyber criminals are targeting smart phones to carry out their malicious activities. Ever since, Android has become the widely used mobile operating system [140], there is significant rise of Android malware [92]. The mode of infection of Android malware is different from that of the PC malware [39]. This poses significant challenges to develop anti-malware solutions as Android malware use completely different types of attack methods for their infection compared to PC malware. The most common methods of infection or sources of Android malware are given below [39].

1. Third party application store: Malware developers may publish malicious applications in third party application store with the name of legitimate applications. If a user downloads these the applications, then the device may get infected with the malware that may steal the personal information of the user or encrypt the files in the device.

2. Man in the middle attack: In this case, connecting an Android mobile device to rogue Wifi hotspots exposes the device to attackers who can send malware to the device.

3. Malvertising: In this technique, a malware is inserted into a legitimate advertisement application. When the user clicks on the advertisement, the device gets infected with the malware. Some malvertisements show up the entire screen of the device and when the user touches the screen, the malicious payload will get downloaded.

4. Scams: In this type of attack, a user is redirected to a malicious web page, using a pop-up or a web redirect. During this process a malware can get downloaded.

5. Physical compromise: In this type of attack, the attackers have physical access to the device. They connect the device into the PC and directly downloads the malicious application.

2.2   Trends in Malware

Because of the widespread usage of connected devices to share data, attackers are now utilizing the vulnerability of the network, operating system and the software to launch malicious attacks. According to a survey of National Institute of Standards and Technology (NIST)[43], the number of vulnerabilities showed a drastic increase in the year 2020. Despite developing and publishing patches, the attackers are finding new ways to evade various attack detection mechanisms.

The latest trend shows that malware come equipped with multiple functionalities with the help of plugin interfaces[162]. The PlugX malware is one such malware[51].

The plugins allow the malware to update its functionality without the need for reinstalling the malware. Further, these plugins allow the malware to adapt to the changes in the new environment. That is, if the environment uses an antivirus, the malware can hide its malicious functionality. In the past, malware attacks were not targeted on a specific user or organization. However, the new trend shows that 70–90% of the recent malware attacks were targeted on specific organizations. Recent trend is the emergence of polymorphic malware to evade the detection by antivirus software. Another design trend is the type of evasion techniques used to hide the presence of the malware. It is seen that the recent malware use a variety of evasion techniques like dead code insertion, packing, anti-emulation technique, etc. for hiding its behavior.

2.2.1 Trends in Windows malware

In the case of Windows malware, file-less attacks have emerged in which a malware utilizes legitimate program for infection[144]. File-less malware can evade the most advanced threat detection mechanisms[149]. File-less infection go straight into the memory and operates in it. Many file-less malware utilize the Windows Powershell, a tool used for configuration management and task automation for carrying out the attack. These attacks often employ social engineering to click an attachment or a link in the phishing e-mail. Besides file-less malware, ransomware and cryptocurrency miners are also on the rise[184][135]. According to the report issued by FireEye[201], the ransomware related breaches showed a dramatic increase by the year 2020. Further, most of the attacks targeted manufacturing organizations. The ransomware also targeted academic institutions when the universities started their classes virtually due to the outbreak of COVID-19[48]. The ransomware attacks pose serious threat when the attackers steal sensitive information before encrypting the device. In this type of ransomware attack, the victims are threatened to make sensitive information public if they fail to pay the ransom[201]. Another type of ransomware attack called Ransomware as a Service (RaaS) attack [47] also emerged. In RaaS attack, the malware developers sell out malware kits to individuals who need to launch attack against an enterprise. These kits are sold in the dark web[147].

2.2.2 Trends in Android malware

In the case of Android, the Google Playstore detection mechanism is not at all effective in detecting malicious applications. The recent trend in Android malware shows a drastic increase in the botnet attacks[44]. A botnet is a group of mobile devices that are controlled by the attackers without the knowledge of the users[64]. According to the threat report of Mcafee, a new type of Android malware called Shopper[45] exists. This malware can take the users Google or Facebook account credentials and then post reviews on the popular entertainment and shopping sites on behalf of the user. For evasion, these malware use hide icon method. This malware is distributed through malvertising and it is found in the Discord chat service[19]. Another trend observed is the type of evasion methods adopted. Nowadays, most of the malware use self-hiding mechanism to hide their functionality[180]. These malware can masquerade as a legitimate application with the similar icons and name and can hide their icon soon after installation. A fakeapp malware is one such malware that mimics legitimate FaceApp Android application and accesses the gallery of the user[23].

The most dangerous trend observed in Android malware is that some malware have been found to spy and collect sensitive information about a country's political and defense matters. MalBus is one such malware[37] that was developed by a South Korean developer. The malware pushes a malicious payload and then uploads it to the Google Play account. When a user downloads it, the malware scans the device for the documents that are related to the military and the state. The applications that are infected by these malware also contain an additional library to download malicious plugins.

Recently researchers have uncovered two novel Android surveillanceware (surveillance malware) families called Hornbill and SunBird used by an advanced persistent threat (APT) group to target military, nuclear and election entities. These two malware families, have sophisticated capabilities to exfiltrate SMS messages, contents of encrypted messaging apps, geolocation, and other types of sensitive information. Hornbill was first detected in early May 2018, and the newer samples of the malware were detected in December 2020. The SunBird was first detected in 2017 and it was last seen active in December 2019. The samples of SunBird was found hosted on third-party app stores, where it was disguised as a security service application. The Hornbill applications impersonate various chat and system applications. Both malware families are able to collect call logs, contacts, device metadata (such as phone numbers, models, details of manufacturer, and version of Android operating system), geolocation, images stored, and WhatsApp files. SunBird has more malicious functionalities than Hornbill, with the ability to upload all data at regular intervals to its servers. Also, it can run arbitrary commands as root or download malicious payloads. In contrast, Hornbill is more like a passive reconnaissance malware than SunBird. Hornbill uploads data only in the initial runs and not at regular intervals like SunBird. After that, it only uploads changes in data.

2.3   Types of Malware Detection Mechanisms

Malware detection is a very important task because a potentially malicious software can damage the devices and the files. The conventional signature based malware detection mechanisms are incapable to detect sophisticated zero day malware. This is because a polymorphic malware can have a different signature that is not stored in the antivirus database. Hence, nowadays many malware detection mechanisms are using machine learning techniques[199]. To detect the malicious behavior of an application, static, dynamic and hybrid mechanisms are used [72][68]. Static mechanisms analyze the application without running the code. The advantages of static analysis is that, it has less overhead and high code coverage. However, the disadvantage is that, these mechanisms fail to detect evasive malware that use dynamic code loading and obfuscation. To deal with these issues, dynamic malware detection mechanisms can be used. In dynamic detection, the malware is made to run in a sandbox environment and the execution traces are gathered to capture the malicious behavior. Despite its advantages, some malware can use anti-emulation techniques to detect whether they are running in a sandbox environment in order to hide their malicious behavior. Anti-emulation techniques are found in many different Android malware families, such as the recent Android Adload adware found in Google Play. To utilize the benefits of static and dynamic malware detection mechanisms, hybrid detection mechanisms are also used. However, malware with advanced evasion capabilities continue to emerge every day. Hence security researchers are trying hard to invent new mechanisms to detect emerging malware.

2.4 Malware Types

Malware applications can be classified into different types based on their malicious functionalities. Some malware application can use the sensitive API calls provided by the application framework while the others can exploit the vulnerabilities of the device for performing malicious activities. The functionalities of malware apps are given below [104].

- Stealing private information: Some malware apps tend to steal private information such as credit card numbers, user login credentials, IMEI code etc. from the device.

- Sending SMS to premium rate numbers: Some malware apps tend to send text messages to the premium rate numbers in the background.

- Surveillance attacks: Some malware apps tend to capture the surroundings by secretly taking the photos or videos in the background.

- Recording phone calls: Some malware apps tend to record the incoming and outgoing phone calls in the background.

- Encrypting the files: Some malware apps tend to encrypt the files in the device and demand ransom for decryption.

- Locking the device: Some malware apps tend to lock the device and demand ransom for unlocking.

- Showing advertisements: Some malware apps tend to show the advertisements in the form of popups for revenue.

The different types of malware applications according to their functionalities are the following:

1. Trojan Spy;
2. Trojan SMS;
3. Backdoor;
4. Ransomware;
5. Exploits; and
6. Botnet.

Trojan spy can record the activities in a device without the knowledge of the user. The spyware can perform various kinds of activities such as keylogging, recording phone calls, stealing SMS, etc. Triout and Bouncing golf are two malware families in this category. Triout was identified in 2018[54]. Triout can record the phone calls and send it to a remote server. Bouncing golf was identified in 2019. Bouncing golf[40] can collect the information such as SMS messages, contacts, list of installed apps, etc. and send them to a remote server.

Trojan SMS can send text messages to premium rate numbers frequently in the background. Boxer and Opfake are two malware families in this category. Boxer was identified in 2011. It sends SMS to premium rate numbers in the background. Opfake was identified in 2012. It also sends SMS to premium rate numbers in the background.

Backdoor can transfer the control of a device to a remote server without the knowledge of the owner. Here, the malicious server can perform various attacks such as stealing information, monitoring user activities and so on. AndroRAT is a backdoor malware family. It was identified in 2013. AndroRAT can control the device and perform various kinds of attacks such as initiating phone call, getting device contacts, etc.

Ransomware is a type of malware that blocks users from accessing the device until he/she pays a ransom to the attacker. Simplocker is a sample ransomware family. It was identified in 2014. Simplocker can encrypt the files in a device and demand ransom for decryption[49].

Exploits tend to utilize the vulnerabilities in the device kernel for performing privileged operations. Gingermaster is a sample malware family in this category. It was identified in 2011[27]. Gingermaster can exploit vulnerabilities in Gingerbird version of Android OS through an exploit code for performing various kinds of privileged operations.

Botnet is a group of infected devices communicating among themselves to perform malicious activities such as DDoS attack. These devices are infected with malware which enable the attacker to control them. Zazdi is an example of a botnet family. It was identified in 2019. Zazdi botnet uses Firebase Cloud Messaging (FCM) services to communicate with the infected devices.

2.5   Malware Attacks in Android

Malware attacks in Android can occur in the following three ways:

- Drive by download attack and
- Update attack and
- Repacking attack.

2.5.1   Drive by download attack

In this case, the attacker first creates a malicious website containing malware applications which can automatically download into the systems. Then the attacker sends the link of the website to the victims via e-mail or SMS messages or social networking sites. These types of malicious websites contain misinformation which prompt some users to visit that page [96]. When a user visits the website, the malware gets downloaded automatically to the phone. For example, Rumms is a drive by download Android malware that is distributed via SMS messages [203]. When a user clicks the link provided in the SMS message, the Rumms malware gets downloaded automatically to the phone.

2.5.2   Update attack

In this case, the malware developer publishes an application in the app stores which updates with malware functionality automatically after installing it. For example, Vmvol malware app gets updated with a malicious payload after installing it [203].

2.5.3   Repacking attack

In this case, an attacker adds a malicious payload to a goodware application. First the attacker takes a legitimate app and decompresses it. In the next step, the byte code is edited . In this step, the attacker uses third party tools to recover the source code from the byte code for inserting the malware code into it and compile the modified code. Finally, the attacker compresses the result and signs with a self-signed certificate. For example, Walkinwat malware application[29] is a repacked version of legitimate Walktext application

2.6   History of Malware Attacks in Android

The first mobile virus called 'Cabir'[3][15] emerged in 2004 and it targeted the Symbian OS running on an ARM processor. When Android OS gained popularity, the first

Android malware called the 'FakePlayer' emerged in 2010[4]. This malware came disguised as a media player and it sent messages to some premium numbers with each messages costing around 5 dollars. When the malware was executed, it displayed messages in Russian language. In 2016, HummingBad malware[62] was found in the wild. It was created by a Chinese advertisement company. This malware also installed fraudulent applications in the infected device.

In 2010, the first trojan called 'Zitmo'[67] emerged. It was designed to steal the mTan codes sent by the banks. Later in 2011, Droiddream[57] a malware that sends information to the remote server was found. In 2012, Boxer[56] malware was seen in the wild. It was distributed through messages and once installed, it sent messages to premium numbers. In the year 2013, FakeDefender[52], the first ransomware that targeted Android application was found. It appeared as a fake antivirus program. The malware displayed a picture of an animal peering out of the letters 'OZ'. In 2014, the SimLocker malware emerged[49]. The malware was a ransomware that scans the SD card of the device and encrypts certain file types. The files that targeted where .mp4, .jpeg and .png. In 2015, Gazon[11], a financial fraudulent application was seen targeting Android. In the year 2017, a malware named ExpensiveWall was found sending messages to premium numbers. Other malware like Marcher, Xavier, DVMap and BankBot also emerged in 2017[2]. Marcher was found stealing the login credentials of the user and it was found inside the third party application sites. This malware displayed fake login pages of Citibank, Walmart, Paypal, etc. Xavier was a trojan adware that records the call, changes the ringtone, etc.[2]. Besides that, this malware used advanced mechanisms to evade static and dynamic analysis. DvMap on the other hand installed the malicious modules and executed it with the root privilege. Bankbot malware/Anubis steals the payment card data with the help of fake overlay screens that looked like legitimate login pages. In the year 2019, Cereberus banking trojan was detected. It had Remote Access Trojan (RAT) [25] capability. In addition to that, the malware also accessed the device's unlock pattern.

With the outbreak of the COVID pandemic in the year 2020, malware developers utilized the functionality of corona virus application tracker to spread the banking malware Anubis[10]. Another version of corona virus app was seen as a spyware taking the photos, recording videos, etc. In the year 2020, Joker malware [66] was also seen in the Google PlayStore. This malware can access the device information, contact list, etc. of the user. The malware was also able to sign up for premium wireless application protocol (WAP) without user's knowledge. In the early 2021, researchers also found a spyware targeting the users in Pakistan[41]. The spyware came hidden in a legitimate application and it exfilterates the sensitive data of the users like the contact list and SMS messages. Despite powerful security mechanisms, new malware continue to emerge day by day. Hence powerful and effective malware detection mechanisms have become a need of the hour.

2.7 Conclusion

In this chapter, we discussed about Android malware, trends in malware, malware types and the importance of malware detection. We also discussed the popular malware attacks in Android. Since malware threats are evolving rapidly, more effective threat detection mechanisms are needed. Since artificial intelligence (AI) powered malware are also on the rise[95], the effectiveness of machine learning techniques in detecting advanced malware threats should also be investigated.

3

Static Malware Detection

Static analysis is the technique of detecting malware applications by analyzing their source code. The source code associated with an application is obtained using reverse engineering tools. The source code contains features like API calls, permissions, hardware components, intents, intent filters and app components. The advantage of static analysis is that it has high code coverage and less analysis overhead. In this chapter, we present various existing static malware detection mechanisms.

3.1 Reverse Engineering and Static Analysis

Reverse engineering is the process of obtaining the source code of an Android application to understand its functionality. Reverse engineering is done for the following reasons.

- To find the vulnerabilities in the application code.

- To detect malicious code for malware analysis.

- To modify the functionality of Android application.

In Android reverse engineering involves the process of disassembling and decompilation. Disassembling refers to the translation of bytes to mnemonics. In Android, smali code is the output obtained after disassembling the Android application.The process of converting the application binaries to the high-level language in which the source code is written is called decompilation. In Android, the decompilation is a process in which the dex files are converted to the java source code files.

3.1.1 Reverse engineering using Apktool and Dex2jar

In this section, we will discuss how to reverse engineer an Android application using Apktool and Dex2jar. The various steps are given below.

- Step 1: Install Apktool.

- Step 2: Download the Android application that we want to analyze and put it in the same folder where apktool is located, as given below.

DOI: 10.1201/9781003121510-3

```
Home@abc   :~$ cd Analysis/
Home@abc   :~/Analysis $ ls
aapt apktool candycorn.apk
Home@abc   :~/Analysis  $
```

- Step 3: Run the command,

```
Home@abc    :~/Analysis $ ./apktool d [app].apk
```

Once the above code is entered, a new directory with the decompiled contents will be created. The directory contains AndroidManifest.xml file and smali code. Smali code is the bytecode version of Java code and it is similar to that of the assembly code. A sample smali code is given below.

```
.class public La;
.super Ljava/lang/Object;
#interfaces
.implements Landroid/text/TextWatcher;
#instance fields
.field final synthetic a:(Lgywwv/jvyjsd/sordvd/ActivityCard);V
```

After disassembling, to analyze the Java source code of the application, we can use dex2jar and JD-GUI. Dex2jar to convert the dex files to jar (java) files. To view the java files we can use JD-GUI. This can be done as follows:

- Download dex2jar.

- Extract the apk.zip and open it.

- Copy classes.dex file from the apk folder and paste it to the dex2jar folder.

- Run the command: sh   d2j-dex2jar.sh   classes.dex to obtain classes_dex2jar.jar file.

- Open the generated classes_dex2jar.jar file using JD-GUI.

3.1.2   Static malware analysis tools

Static Android malware analysis tools are very useful for malware researchers to analyze the malicious functionality of Android application. The popular static Android malware detection tools are given below.

- Amandroid – Amandroid is a static analysis tool that is used for Inter-component data analysis of Android applications.

- APK Analyzer – APk analyzer is a static analysis tool that has the ability to compare two apk's.

- SmaliSCA – SmaliSCA is a static analysis tool for examining the smali files.

- Maldrolyzer – Maldrolyzer is a tool to extract 'actionable' data such as C&Cs, phone numbers, etc. from the Android malware.

- Argus-SAF – Argus-SAF is a tool with different capabilities such as Java native interface analysis, annotation based analysis, etc.

- DroidRA – DroidRA provides reflection analysis for Android malware.

- Androwarn – Androwarn is a Dalvik bytecode analysis tool that can identify suspicious permissions and other activities of Android applications.

- PScout – PScout is an Android API call to permission mapping tool that can be used for malware analysis.

- Androguard – Androguard is powerful tool that can be integrated with other tools for malware analysis. It provides class analysis, method analysis, permission analysis, etc.

3.2 Components of Android Application

An Android application has the following components[5].

- Java Source Files: The Java files contain the application's source code written in Java programming language.

- res/drawable-hdpi: This directory contains values for the resources that are required for running the Android application. It includes style, color, dimensions, etc.

- res/layout: This directory contains details about the application's user interface.

- res/values: This directory contains a collection of resources, like colour definitions and strings.

- AndroidManifest.xml: It stores meta-data such as package name, permissions required, definitions of one or more components like Activities, Services, Broadcast Receivers or Content Providers, minimum and maximum version support, libraries to be linked, etc. In manifest file, the developer can specify the hardware components and software requirements of the application. The developer can specify the minimum and maximum SDK required in the manifest file.

- Build.gradle: This file is an autogenerated file that contains compileSdkVersion, targetSdkVersion, versionCode, versionName, etc. The compile SdkVersion is the version in which the Android application is debugged and compiled. The targetSdk version is the version of Android that the application was developed to run on.

3.3   API Call Analysis

API stands for Application Programming Interface. It can be defined as a set of protocols, procedures, and tools that allow interaction between two applications [65]. In
Android, the application layer which forms the top of the native libraries provides the
API. These (APIs) are built in the form of java classes. The API consists of enormous
class library (a set of packages) suitable for building our own applications. API level
of an Android is an integer value that helps to identify the API revision offered by
a version of the Android platform [13]. The new API level launched each time by
the Android is compatible with the older versions. The code below shows how API
packages are imported in the java source code of an Android application.

```
1.package com.abc.telephonymanager;
2.import os.Bundle;
3.import android.content.Context;
4.import android.view.Menu;
5.import android.app.Activity
6.import android.telephony.TelephonyManager;
7.public class MainActivity extends Activity{
8.setContentView(R.layout.activity_main);
9.TelephonyManager t=(TelephonyManager)
getSystemService(Context.TELEPHONY_SERVICE);
10.String IMEINumber=t.getDeviceId();
11.String subscriberID=t.getDeviceID();
12.Sring SIMSerialNumber=t.getSimSerialNumber();
}
```

In the line 6 of the code above *android.telephony* is the package name and the *TelephonyManager* is the API class name. The API calls are the methods of the API classes.
Lines 10, 11 and 12 show the API methods of the class *TelephonyManager*. The API
classes and methods are very crucial for identifying the behavior of Android application. The smali codes can also be directly analyzed before converting to the java
source files to understand the API's of the application. The code below shows how
the API packages, classes and methods appears in smali. Here *Landroid/telephony* is
the package name, *TelephonyManager* is the API class name and *listen ()* is the API
method name.

```
Landroid/telephony/TelephonyManager;>listen(Landroid/telephony/
PhoneStateListener;I)V
```

Android malware may use certain API's to access the sensitive resources such as
location, device Id, etc. of the mobile phone to perform its malicious activities. The
following section describes how API can be used to detect Android malware.

3.3.1 API's used by malware applications

We can classify different API's by the type of requested utilities and resources to identify the malicious behavior [68]. They are application specific resources API's, Android framework resources API's, DVM related resources API, system resources API's, and utilities API's.

1. **Application Specific Resources API's**

   The application specific resources API's can be classified into Content resolver class, Context class and Intent class.

   (a) Content resolver class: The content resolver class gives access to content providers. Malware may use methods such as *delete()*, *insert()*, *query()*, etc. of this class to carry out the malicious activities.

   (b) Context class: Context class provides information on classes, resources and assets. The *startService()* method in this class is frequently used by the malware to start a service in the background. The methods like *getFilesDir()* and *openFileOuput( )*, etc. are other API's of this class used by the malware to create files and to obtain their absolute paths.

   (c) Intent class: Intents are used to interact with the phone's hardware and also to launch activities and services. The frequent APIs invoked by malware in this class are *setFlags(), addFlags()* and *setDataAndType()*.

2. **Android Framework Resources API's**

   The Android framework resources API's can be classified into Activity Manager class, Package manager class, Telephony SMS manager and Telephony class.

   (a) ActivityManager class: The ActivityManager class helps to interact with other activities running in the device. The *getRunningServices()* method invoked by the malware is used to find whether there are any antivirus services running in the device. Another method called *getMemoryInfo()* is used by the malware to check whether there is enough memory for background process to kill other processes. The method *restartPackage()* is invoked by the malware to kill other application services.

   (b) PackageManager class: The package manager class contains information about the application packages installed on the device. The *getInstalledPackages()* method is used to scan the device to check whether there are any antivirus programs running in the device to kill it.

   (c) Telephony/SmsManager & Telephony/gsm/SmsManager: The methods in these classes allow malware developers to send premium rate SMS from infected devices thereby causing financial losses to the users. The *sendTextMessage()* is one such method of this class.

(d) Telephony Manager: Malware may use methods of this class such as *getSubscriberId( )*, *getSimSerialNumber( )*, *getLine1Number( )*, *getNetworkOperator( )*, *getCellLocation( )*, *getDeviceId( )*, *getNetworkType( )*, etc. to access the device information of the user. Android malware such as Ginmaster[58], Exodus[22], etc. used this API.

3. **DVM Related Resources API's**

   The DVM related resources API's can be classified into DexClassLoader class and Runtime and System class.

   (a) DexClassLoader class: The DexClassloader class helps to load a classes.dex file. Malware may use the method called *loadClass( )* to execute the code that is not a part of the application so as to evade the malware detection mechanisms.

   (b) Runtime and System class: In this case, the method called *Runtime.getRuntime.exec ( )* in the Runtime and System class helps the malware developers to execute malicious code in the form of shell scripts to evade detection. The method *loadLibrary( )* is used to dynamically load the native libraries. This can be used by the malware developers to run native codes.

4. **System Resources API's**

   The system resources API's can be classified into ConnectivityManager, NetworkInfo and WifiManager class, HttpURLConnection and Sockets class and OS package and IO package class.

   (a) ConnectivityManager, NetworkInfo, and WifiManager class: The ConnectivityManager, NetworkInfo and Wifi Manager classes provide network related functionalities such as answering queries about different connections and network interfaces. Android malware may call API's within ConnectivityManager class such as *getNetworkInfo( )*, *NetworkInfo( )*, *getExtraInfo( )*, *getTypeName( )*, *isConnected( )*, *getState( )*, and the methods in WifiManager class such as *setWifiEnabled( )* and *getWifiState( )* to establish a network connection and interact with malicious remote servers. This API's were used by the Android malware Basebridge [55] used these API's.

   (b) HttpURLConnection and Sockets class: The methods used in these classes are used for establishing connections to a remote server and to send or receive data over the web. *SetRequestMethod( )* and *getOutputStream( )* etc. are the methods used by the malware to establish a connection with a remote server.

   (c) OS package class: The method such as *kill process( )* of the OS package class is often invoked by the malware to kill the process that is running on a given process id. The OS package also contains several API for process and thread management.

(d) IO Package class: This package provides read and write to datas-treams, files, etc. Malware uses API's such as *writeBytes()* to upload certain files to a malicious url, *readLines()* to read malicious payload and *delete(), mkdir()* API's for launching a variety of malicious activities.

5. **Utilities API's**

The utilities API's can be classified into string and String Builder class, Crypto class, ZipInputStream class and w3c.dom class.

(a) String and StringBuilder class: The methods in this class are widely employed for code obfuscation by the malware. API calls such as *indexof(), getBytes()* and *replaceAll()* are used for creating and manipulating the strings.

(b) Timer: Malware may use APIs of this class such as *schedule()* and *cancel()* to evade the detection.

(c) Crypto class: The methods in this class allow implementing cryptographic operations. API calls such as getInstance(), doFinal() and Crypto.spec.DESKeySpec() are used by the malware for code obfuscation.

(d) ZipInputStream class: Malware may use methods to decompress the malicious .zip/.rar files. Read(), close(), getNextEntry(), etc. are other API methods of this class that are used by the malware.

(e) w3c.dom class: Malware may use getDocumentElement(), getElementByTagName(), and getAttribute() for communicating with a bot and encode the data.

3.4 API Call-Based Static Detection

There are two types of API call-based malware detection mechanisms. They are mechanisms that use the independent occurrence of API and mechanisms that use API call graphs. In this section, we explain some API call-based malware detection mechanisms.

3.4.1 Mechanisms using the independent occurrence of API

In this mechanism, the independent occurrence of API call-based features are analyzed for finding the malicious behavior. It is computationally less complex to construct the binary feature vector of an application based on independent occurrences of API call-based features. The following mechanisms use independent occurrence of API for detecting malware.

In DroidAPIMiner [68], the source code API level features such as package name, class name and function name are analyzed using machine learning algorithms for finding malicious behavior. In this mechanism, they selected the API level features which are more frequently found in malware apps in their dataset. Sometimes goodware apps may use non malicious features which are frequently found in malware apps. In such cases, a goodware app might get misjudged as a malware.

In [158], API calls are used as feature vectors for a trained deep Convolutional Neural Network (CNN) classifier for predicting whether the application is a malware or not. The authors used a very limited number of malware (216 samples) to evaluate their approach.

In [123], Hou et al. proposed a mechanism to detect Android malware apps by analyzing the API calls in it. In this approach, a trained Deep Belief Network (DBN) classifier is used to predict whether the application is a malware or not based on the API calls in it. This approach is computationally expensive.

Han et al.[117] proposed an Android malware detection mechanism using API calls. Their mechanism used SVM for detecting malware. Shankarpanni et al. [181] proposed a malware detection mechanism using API calls to detect obfuscated malware. In [142], the authors proposed a detection mechanism that used structural and behavioral features of API to detect malware. In [69], the authors proposed a malware detection mechanism by combining permissions and API calls. Android tend to revise its API calls time to time. Hence, evolving malware apps can use newer API calls for performing malicious activities. In such cases, newer benign applications may get wrongly flagged off as malicious and vice versa in these detection mechanisms. Hence, frequent classifier retraining is required in these mechanisms.

3.4.2   Mechanisms Using API Call Graphs

In these mechanisms, the API call or inter component communication graphs are analyzed for malware detection. From the API call sequence or graphs, we can easily infer the underlying semantics behind a family/category of Android malware. The following works use API call graphs for detecting malware

Zhang et al. [218] proposed a mechanism to detect Android malicious apps from its API call dependence graph for finding whether the application is malicious or not. In this approach, the graph similarity metrics are used for measuring the similarity between API call dependency graph associated with the application and those of known malicious applications. This approach cannot detect malware apps having unseen API call sequence.

In Faldroid [102], frequent API call subgraphs generated by the malware families are used for finding malicious behavior. This approach has two steps. In the first step, authors employed a clustering based approach to extract common malicious behaviors (frequent subgraphs) of each malware family. In the second phase, they employed a weighted sensitive API call-based graph matching approach to classify the unknown malware applications. This approach cannot detect malware apps which contain privilege escalation exploits.

3.5 Permission and Intent-Based Static Detection

In this section, we explain how permissions and intent analysis are used for Android malware detection.

3.5.1 Permission analysis

A permission is a unique text string that is defined by Android or third party developers [103]. In Android, permissions are requested when an application is accessing the resources of the device. In Android, the permissions are divided into four types based on the protection levels they offers. They are normal, dangerous, signature, and signatureOrSystem[63]. Types of some sample permissions are given in Table 3.1. These levels are assigned based on the damage they may cause if the user grants it. While installing the application, the Android package installer will not ask the user for approval for the permissions that have the safe protection levels. But in the case of the permissions with dangerous protection levels, the user will be asked for a consent to install the applications that needs it. The Android system grants signature permissions at the application installation time provided that the app that is asking for the permission must be signed with the same certificate as that of the application that defines the permission. The signatureOrSystem level permission on the other hand requires the application to be a system application. However, this type of signatureOrSystem level permission is deprecated from API level 23 onwards.

TABLE 3.1 Various types of permissions.

Type of Permission	Example
Normal	ACCESS_LOCATION_EXTRA_COMMANDS ACCESS_NETWORK_STATE CHANGE_NETWORK_STATE ACCESS_WIFI_STATE
Signature	BIND_ACCESSIBILITY BIND_ACCESSIBILITY_SERVICE, BIND_AUTOFILL_SERVICE, BIND_CARRIER_SERVICES
Dangerous	READ_CONTACTS WRITE_CONTACTS GET_ACCOUNTS ACCESS_FINE _LOCATION ACCESS_COARSE_LOCATION SEND _SMS RECEIVE_SMS
SignatureOrSystem	DELETE_PACKAGES

Android permissions prevent malware developers from accessing the sensitive resources of the device. However, the vulnerability in the permission framework of Android can be exploited by the malware to execute various malicious activities. In 2019, researchers found a vulnerability called "StrandHogg" that was able to exploit the permission pop up mechanism in Android. In this type of attack, malware abused the permission called SYSTEM_ALERT_WINDOW, that allowed an application to display a window on top of other applications. By using this pop up mechanism, the malware displayed fake banking login pages and fraudulent ads. Initially, Google blocked the apps that used this permission. But, it affected the normal functionality of legitimate apps such as Facebook that need to show pop up windows for chat operations. Hence from Android 6.0 onwards, the restriction has been taken out by the Google. However, the applications that use this permissions are analyzed by the Bouncer mechanism before making it available to the users. According to the recent threat report[30], 74% of ransomware, 57% of adware and 14% of banker malware abuse this permission for doing malicious activities. There are several other permissions that are used by the malware for their activities. The following subsection provides some of the permissions abused by the malware developers to launch attack against Android.

3.5.1.1   Permissions used by the malware applications

The various permissions used by the malware applications and the details are given below.

1. ACCESS_COARSE_LOCATION and ACCESS_FINE_LOCATION: These permissions are used by the malware applications to locate the network provider information and GPS information of the user. Tekya malware [169] and Malbus malware[38] use this permission to access the GPS location and network provider information.

2. ACCESS_NETWORK _STATE: Malware use this permission to monitor the network connection of the device. This permission is used by a malware to connect the device to a malicious remote server. DroidKungfu malware uses this permission to collect the information about the device[20].

3. ACCESS_WIFI _STATE: This permission is used by a malware to access the wifi connectivity of the user. Malware may send wifi information to the remote server to monitor the wifi connectivity of the user. Corona-Tracker malware that appeared recently uses this permission to obtain the wifi connectivity information[17].

4. BROADCAST_PACKAGE_REMOVED: This permission is used to notify that an application package has been removed. Malware may use this to kill other running applications such as antivirus applications.

5. CALL_PHONE: This permission helps to make calls without user's consent. Malicious applications use this permission to make calls resulting in huge call charges to the user. Basebridge [55] malware uses this permission to make phone calls.

6. CHANGE_WIFI_STATE: This permission is used to change the wifi state of the device. This permission is used by the malware application to connect to the malicious servers. FakeAngry [24] malware uses this permission.

7. DELETE_PACKAGES: This permission is used to delete the application packages that are installed in the device. Malware may use this to delete the antivirus programs that are running on the device.

8. INSTALL_PACKAGES: This permission is used to install packages to the device. Malware may use this to install malicious application packages.

9. INTERNET: This permission is used by the application to access the Internet. Malware may use this permission to establish a connection with a remote server. The Cerebrus malware [25] uses this permission.

10. KILL BACKGROUND PROCESSES: This permission is used by the application to kill the background process. This permission may be used by the Android application to kill the antivirus process running in the background.

11. MOUNT_UNMOUNT_FILESYSTEMS: This permission is used by the Android application for mounting or unmounting storage devices. Malware may use this permission to access the SD cards in the device for reading and writing the data from or to the device.

12. PROCESS_OUTGOING_CALLS: An application with this permission is able to monitor the details of outgoing calls. A malware may use this permission to monitor the phone numbers and personal details that are stored in the device.

13. READ_CONTACTS: This permission allows an application to read the contact information saved in the device. Malware may use this information to infect other devices using the contact information. Malware such as Comebot [16] and Cerebrus [25] use this permission.

14. READ_PHONE_STATE: This permission allows to read the current cellular network information and the status of ongoing calls from the device. Malware may use this permission to eavesdrop the phone call activities of the user. Malware such as Comebot[16], and Exodus[22] use this permission.

15. READ_SMS: This permission is used by the application to read SMS. Malware may use this permission to gather the sensitive information of the user. Basebridge[55] malware uses this permission.

16. RECEIVE_BOOT_COMPLETED: This permission allows the application to start as soon as the system has finished booting. Malware may use this permission to start the malicious activity as soon as the system is booted. The malware DroidKungfu[20] uses this permission.

17. RECEIVE_SMS: This permission is used by the application to receive SMS for legitimate services. This permission is used by the malware for

listening and forwarding the SMS's to the remote server without the user's knowledge. FakeInstaller[1] malware uses this permission.

18. SEND_SMS: This permission allows an application to send SMS. This permission is used by the malware to send SMS to premium rate numbers. Malware such as Mobts[16], FakeAngry[24] and FakeInstaller[1] use this permission.

19. SET_WALLPAPER: This permission that is used to set the wall paper is used by a malware to download malicious content from the remote server. Certain malware may also use this permission to crash the device by re-booting it many times.

20. WAKE_LOCK: This is a permission that is used to keep the device's screen active. Malware may use this permission to perform malicious activities in the background by keeping the device active.

21. WRITE_APN _SETTINGS: This permission allows the application to write the APN settings and to read the sensitive fields like username and password. Malware may use this permission to gather sensitive information of the device.

22. WRITE_CONTACTS: This permission allows an application to write the user's contacts data. Malware application may use this to access the accounts like Google, Facebook, etc. The malware Mobts [16] uses this permission.

23. WRITE_EXTERNAL_STORAGE: This permission is used to read or manipulate the files stored in a phone. Malware may use this to access the files stored in a device. Malware Ginmaster[58] uses this permission.

24. WRITE_SMS: This permission is used to write SMS. Malware may use this to send SMS to the premium rate numbers. Malware such as FakeAV[59], and Dendroid[18] use this permission.

Table 3.2 shows the permissions used by some of the popular malware applications for doing various malicious activities.

3.5.1.2 Component-based permission escalation attack

In Android, every application runs in its own sandbox.This means that every application has its own resources and neither of the applications can interact with each other. However, a transitive permission attack can break the Android's security model by allowing an unprivilaged application to utilize the privileges of another application [205]. A component based permission escalation attack is one such attack. This attack exploits the MAC reference model. Figure 3.1 shows the component-based permission escalation attack.

Here the three android application *A*, *B* and *C* are isolated from each other. Each of the applications are running in their own sandboxes. Application *A* has two components *A*1 and *A*2 with no granted permissions and application *B* has two components *B*1 and *B*2 with the permission *P*2. Let the application *C* has two components *C*1

TABLE 3.2 Permissions used by malware families.

Permission	Geinimi	DroidDream	PjApp	GoldDream	Adrd
ACCESS_COARSE_LOCATION	X			X	
ACCESS_FINE_LOCATION	X			X	
ACCESS_NETWORK_STATE				X	X
ACCESS _WIFI _STATE				X	
BROADCAST_PACKAGE_REMOVED				X	
CALL_PHONE	X			X	
CHANGE_WIFI_STATE		X		X	
DELETE_PACKAGES				X	
INSTALL_PACKAGES				X	
INTERNET	X	X	X	X	X
KILL_BACKGROUND_PROCESSES					
MOUNT _UNMOUNT_FILESYSTEMS	X				
PROCESS _OUTGOING_CALLS	X			X	
READ_CONTACTS	X				
READ_PHONE _STATE	X	X		X	
READ_SMS				X	
RECEIVE_BOOT_COMPLETED				X	X
RECEIVE _SMS			X	X	
SEND _SMS	X				
SET_WALLPAPER	X				
WAKE _LOCK					
WRITE_CONTACTS	X				
WRITE_EXTERNAL_STORAGE	X				

and $C2$ which are protected with permission $P1$ and $P2$. The application A's components are not allowed to access the components of application C directly because they are protected by permission $P1$ and $P2$. However, application B is allowed to access the components of application A. In component based permission escalation attack, the application A accesses the components of application C indirectly through the components of B.

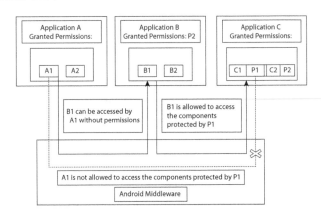

FIGURE 3.1 Component-based permission escalation.

3.5.2   Intent-based analysis

In this section, we explain how intents in Android application can be used for malware detection. We also explain the intent-based vulnerabilities in Android and how the intents are used by the malware applications to launch malicious activities.

Intents are requests that are given to the activities, external applications and built-in Android services. An application has many activities and each of the activities may have buttons, texts and labels associated with it. The intent mechanism helps to push data from one activity to the other. Intents are defined in the AndroidManifest.xml file in the form of intent filters. An intent object carries the information to determine the component to start and the actions for the components. An intent object consists of the following.

1. Component Name: The intent object holds the name of the component of the Android application. The component name indicates the activity, service or BroadcastReceiver class.

2. Action: Actions are used to define the task that is to be performed by the components.

3. Data: The data defines the type of data that is specified by the intents.

4. Category: The category is a string that is used to specify the type of component that should handle the intent.

5. Extras: The extra values are used to provide additional information to the components.

6. Flag: The flags are used to tell the Android system how to start an activity, and how the activity is carried out after it is launched.

There are two types of intents. They are explicit intents and implicit intents. Implicit intents used to perform an action based on specific data or value. For example, if the application needs to make a telephone call, the implicit intent would be like,

```
Intent callIntent = new Intent(Intent.ACTION_CALL);
callIntent.setData(Uri.parse("tel:78654989"));
startActivity(callIntent);
```

The explicit intent on the other hand is used for calling one activity in response to a user action or another activity. For instance if the application needs to send a message when the user clicks a button, we use an explicit intent. The explicit intent would be like,

```
public void launchComposeView()
Intent i = new Intent(ActivityOne.this, ActivityTwo.class);
startActivity(i);
```

3.5.2.1   Intents used for malware attacks

Various intents used by the malware to launch malicious activities are given below.

1. android.intent.action.BOOT_COMPLETED: This intent is used by the malware applications to start the malicious activities when the device has finished the booting process. DroidKungfu [20] malware uses this intent.

2. android.intent.action.BATTERY_LOW, android.intent.action.BATTERY _OKAY: Malware may use these intents to check the status of the battery level of the device to load various malicious payloads.

3. android.intent.action.INPUT_METHOD_CHANGED: This intent is used to check whether the device's input method is changed.

4. android.intent.action.ACTION_POWER_CONNECTED: This intent is used by the malware to check whether the device is connected with the power to initiate a malicious app update. The malware Basebridge [55] uses this intent.

5. android.provider.Telephony.SMS_RECEIVED: This intent is used by the malware application to check whether the device has received SMS so as to forward it to a malicious server.

6. android.intent.action.USER_PRESENT: This intent is used by the malware application to check whether the user is present or not by checking whether the device is unlocked. Tekya malware uses this intent[169].

7. android.intent.action.SCREEN_ON: This intent is used by the malware to check whether the device's screen is on.

8. android.intent.action.SCREEN_OFF: This intent is used to check whether the device's screen is off. Malware FakeAV[59] uses this intent.

9. android.intent.action.SIG_STR: This intent is used by the malware application to check the signal strength of the device.

10. android.intent.action.PACKAGE_INSTALL: This intent is used by the malware to check whether the malicious package is installed or not.

11. android.intent.action.PACKAGE_ADDED: This intent is used by the malware application to check whether the malicious package is added or not.

12. android.intent.action.PHONE_STATE: This intent is used by the malware to check whether there are any incoming calls in the device.

13. android.app.action.DEVICE_ADMIN_ENABLED: This intent is used to check whether the administrator privilege is enabled to manage the device.

14. android.intent.action.ACTION_EXTERNAL_APPLICATIONS_AVAIL ABLE: This intent is used by the malware to start the application when it is present in the SD card.

15. android.intent.action.QUICKBOOT_POWERON: This intent is used by the malware to start a service when the device is booted. Comebot malware uses this intent[16].

3.5.2.2 Intent-based vulnerabilities

The recent trends in Android malware show that a malware can exploit the vulnerabilities present in the intent mechanism of the Android[31]. In this section, we discuss about the intent-based vulnerabilities in Android applications.

An Android application may consists many private components which other applications are not allowed to access. However an intent redirection vulnerability allows the malware to access the private components of an application and steal the sensitive data. In the year 2021, thousands of shopping applications were exploited by the intent redirection vulnerability [81]. In the next paragraph, we explain the intent redirection vulnerability in Android.

Android application consists of components such as Activity, Service, Broadcast Receiver and Content Providers. These components are private and can be accessed by other components of the same application. However, if the components are declared as public they can be accessed by the components of other applications also. These components are invoked with intents and specific API's. Suppose there are two applications A1 and A2 in which A1 is a malicious application and A2 is a legitimate application. Let the application A2 has two components a public component denoted by pu and a private component denoted by pr. The private component pr of A2 is designed in such a way to extract and process the data sent from the intent I of the application A1. Here the malware application A1 accesses the private component of A2 as follows. At first A1 sends the intent I to access the public component pu of A2. In addition to I the malware also adds another intent object I' to I to access the private component pr of the legitimate application. When the public component pu receives the intent I, pu extracts and forwards it to pr. The private component pr extracts the data from its intent I which consists of an additional intent object to access the private component pr. In this way, the attacker is able access the private component of the legitimate application. Figure 3.2 shows how this attack is carried out.

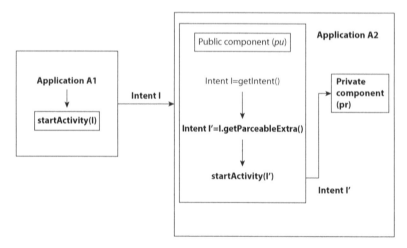

FIGURE 3.2   Intent redirection attack.

3.5.3 Malware detection using permissions and intents

This subsection explains the mechanisms based on permissions and intents for detecting malware.

Features of Android manifest file such as intents and permissions may not change from time to time. Hence, frequent classifier retraining is not required if we are using permissions and intents.

Talha et al. [193] suggested a signature-based malware identification mechanism called Apkauditor. This mechanism analyzes the signatures based on the permissions requested by the application for identifying whether it is malicious or not. The accuracy of this approach was low as compared to other approaches.

Shahriar et al. [179] developed a mechanism to detect Android malware application by analyzing their requested permissions. In the first step, they used Latent Semantic Indexing (LSI) to find the frequent permissions requested by most of the malware applications. In the second step, they searched for these frequent permissions in unknown applications for identifying the malicious behavior. The authors did not mention the accuracy of their approach.

In ICCDetector [213], inter component communication patterns were analyzed for finding the malicious behavior. This approach had two steps. In the first step, a machine learning classifier was trained with the ICC related features of several known goodware and malware apps. In the next phase, a classifier was used to predict whether the unknown application is a malware or not based on the ICC patterns.

Intent-based mechanisms are very accurate in detecting malware apps which perform malicious behavior by invoking the components of other applications. However, these mechanisms will not work in the case of repacked malware apps. That is, an attacker can add malicious functionality without adding or removing the components of an application. In such cases, a malware application may get falsely flagged off as a goodware.

Feizollah et al.[105] developed a mechanism to detect Android malware by analyzing their permissions and intent related features. In the first step, they used Bayesian network to build the malware detection model. In the second step, they used heuristic search algorithms such as K2 and HillClimber to reduce the complexity of the model. The Bayesian network model was able to predict whether an application was malware or not based on the permissions and intents. In this approach, the time required for training the Bayesian network algorithm was very high. In [129] and [124], the authors proposed a malware detection mechanism by combining permissions and intents. Sato et al.[177] proposed a light weight malware detection mechanism using various static features such as permissions, intents, intent filters, number of redefined permissions, etc. Their mechanism gave 91% accuracy. Andre et al.[108] proposed a malware detection mechanism with permissions and intents. Their mechanism gave 97% accuracy with Adaboost Random Forest (ARF) classifier.

Eventhough permissions are useful for detecting malware, certain goodware apps may request dangerous permissions for legitimate reasons. For example, a banking application request SEND_SMS permission for legitimate purposes. In such cases, the goodware app might get misjudged as a malware. Furthermore, malware applications

have tendency to share their user-id with legitimate apps to inherit permissions [204].
All of these factors may degrade the accuracy of the classifier.

3.6   Opcode-Based Static Detection

When an Android application is compiled, Dalvik opcodes are generated. The Android RunTime Environment (ART) translates opcodes into the instruction set of the
processor. There are 221 unique opcodes in an Android application. An example of
Android Dalvik opcodes is given below.

```
add-int/2addr
check-cast
const/4
const-string
goto
if-eqz
if-ne
if-nez
iget
iget-object
```

The opcodes provide the execution paths of Android application and it helps to
understand the functionality of application. The main advantage of using opcodes as
features for Android malware detection is that they can be used to detect malware
without expert analysis[91]. The shell script to extract opcodes from an Android application is give in the Appendix.

3.6.1   Malware detection using opcodes

Analysis of opcodes have been found to be very effective in detecting Android malware. Table 3.3 provides the top 5 opcode 3−grams used for detecting malware. In
this section we discuss about various malware detection mechanisms that use opcode
analysis.

The opcodes are found to be useful features for detecting Android malware since
they are closely related to the application code[91]. Canfora et al.[91] proposed a
technique based on the frequencies of opcode *n*-grams for detecting Android malware. Jerome et al.[125] also proposed a technique based on opcode *n*-grams using
ML techniques. Chen et al.[94] proposed a mechanism called TinyDroid, that uses a
compression technique to reduce the feature set. Their technique was able to achieve
a higher accuracy value with low false alarm rate. Pektas et al. [164] also proposed a
technique using deep learning and instruction call graph for detecting Android malware. Bacci et al. [78] proposed a hybrid mechanism to detect malware by combining

TABLE 3.3 Top 5 opcode 3–grams used by malware applications.

Sl. No	Opcodes
1	move-result-object,move-object/from16,invoke-virtual
2	move-object/from16,iget-object, invoke-virtual
3	if-eqz,check-cast,goto
4	move-object, check-cast, if-eqz
5	move-object/from16,iget-object, move-object/from16

opcodes and system calls to detect malware. Bai et al. [79] proposed a mechanism to detect Android malware by combining Dalvik opcodes and permissions. Kim et al.[131] proposed a multimodel deeplearning technique to detect Android malware by combining opcodes and various features. Bakhshinejad et al.[80] also proposed a mechanism to detect Android malware by compression mechanisms with opcodes as features. In [137], Rao et al. proposed an opcode based malware detection mechanism. Their mechanism used Recurrent Neural Networks (RNN) and gave 96% accuracy. Tang et al. [195] proposed MGOPDroid to detect obfuscated variants of Android malware with opcode sequences as features. In [146], the authors proposed a model based on PercieverIO that can process long sequence of opcodes effectively for malware detection.

Many of the above detection mechanisms can be evaded by a code injection attack[127] [95] [139] that can degrade the accuracy of the detection mechanism.

3.7 Conclusion

In this chapter, we discussed the static analysis approaches used for Android malware detection. We also discussed how the sensitive API's, permissions and intents are used by malware for doing various malicious activities. This chapter will help the cyber security researchers and developers to understand how static analysis mechanisms can be used effectively for malware detection and to develop more effective detection mechanisms.

4

Dynamic and Hybrid Malware Detection

All the static analysis mechanisms can be evaded by dynamic code loading attacks [171]. A dynamic code loading attack consists of three steps. In the first step, a malware developer creates an application with two byte codes: a malicious code and a legitimate code. In the next step, he encrypts the malicious byte code. In the final step, he writes a decryption program in the main byte code (legitimate byte code) to decrypt the malicious byte code and compile the application. Hence, the malicious byte code is not available in plain text for analysis. During the runtime, the malicious code gets decrypted and executed in the system. Hence, it is required to consider runtime information of an application to detect the malware apps which employ dynamic code loading attacks.

4.1 Emulator-Based Dynamic Analysis

Dynamic analysis rely on the runtime features such as network packet level information, system metrics, sensitive API calls and system calls for identifying the malicious behavior. In emulator based dynamic analysis, the dynamic analysis is conducted by executing the application in an emulator. For conducting dynamic analysis, we initially set up an Android emulator in our PC. The steps are given below:

- Download and install the Android studio;

- Open Android studio;

- Go to tools -> Android -> AVD Manager -> Create Virtual Device;

- Create AVD by selecting phone type and API level information and

- Launch AVD (Android Virtual Device) from AVD manager.

After setting up the emulator, we install the testing app into the emulator. The command is given below.

```
adb install <path of the application>   \index{ADB}
```

During the dynamic analysis, we inject several pseudo random events with automated test case generation tools for achieving better code coverage. Here, we used monkeyrunner as automated test case generation for injecting pseudo random events. The command is given below.

```
adb shell monkey -p <package-name> -v <number of events>
```

During the execution of the application, we can use system utilities in the emulator for collecting dynamic information. For collecting system calls, we can use strace tool in the emulator. The command for collecting system calls is given below.

```
strace -p <process-id> -o <output-filename>
```

For collecting the network packets, we can use tcpdump command in the emulator. The command for collecting network packets is given below.

```
tcpdump -i <interface> -w <output-filename>
```

By this way, we can use tools or frameworks in the emulator to collect various other dynamic information such as CPU/memory usages, battery utilization, API calls, etc..

4.2   Dynamic Malware Detection Mechanisms

We classify the existing mechanisms into the following categories:

1. System Metric Analysis (Category 1);
2. Network Packet Analysis (Category 2);
3. Sensitive API Call Analysis (Category 3);
4. System Call Analysis (Category 4).

The comparative analysis of existing dynamic analysis mechanisms are given in Table 4.1.

4.2.1   System metric and traffic analysis (Category 1)

In these mechanisms, the system metrics such as CPU usage, memory usage, battery level information and network traffic are analyzed for malware detection. In this approach, since the system level features may not change from time to time, frequent classifier retraining is not required.

In Andromaly [178], the system level features such as CPU as well as memory usage and battery utilization are recorded while running the application. Then, these features are used in a trained ML classifier for malware detection. The accuracy of this approach is low as compared to other mechanisms.

In [182], the system level features such as CPU usage, memory usage and network traffic information are recorded in every 5 seconds and these features are used in a trained ML classifier for finding the malicious behavior. This approach is tested in

TABLE 4.1 Comparative analysis of dynamic malware detection mechanisms.

Category	Performance	Limitations
Category 1 [152]	Precision: 0.84, F1 Score: 0.84	High false positive rate
Category 1 [182]	Accuracy: 0.99 TPR: 0.98 FPR: 0.01	Performance not clear
Category 1 [178]	Detection Accuracy: 0.88 TPR: 0.91	Low accuracy
Category 2 [82]	Detection Rate: 0.77	High false positives
Category 2 [130]	TPR: 0.96, Accuracy: 0.96	Prone to API call injection attack
Category 3 [119]	Detection Rate: 0.73, FPR: 0.04	Malware detection rate is low
Category 3 [73]	Detection Rate: 0.90	High feature extraction cost
Category 4[89]	N/A	Performance results are not given
Category 4 [71]	Accuracy: 1	Performance not clear
Category 4[199]	TPR: 0.99, Accuracy: 0.99	System call relationship is missing
Category 4[91]	TPR: 0.97, Accuracy: 0.97	Cannot detect unknown malware families
Category 4 [100]	Accuracy: 0.96	High dimensionality of feature vector
Category 4 [122]	Accuracy: 0.92	Require a large training dataset
Category 4[209]	TPR: 0.98 F1 Score: 0.98	High dimensionality of feature vector
Category 4 [85]	Accuracy: 0.94	Feature extraction cost is high
Category 4[208]	TPR: 0.98 F1 Score: 0.98	High dimensionality of feature vector
Category 4[219]	TPR: 0.97 Accuracy: 0.98	High dimensionality of feature vector
Category 4 [83]	TPR: 0.94 F1 Score: 0.91	Requires multiple system call logs
Category 4 [210]	Precision: 0.91 Accuracy: 0.94	Prone to system call reordering attack
Category 4 [99]	TPR: 0.96, Accuracy: 0.94	Emulator customization is required

very limited number of applications (20 goodware and 6 malware apps). Hence the performance is not clear on large and realistic datasets.

In [152], the relevant features related to CPU and memory usages are used as input features of a machine learning classifier for identifying whether an application is malicious or not. In this approach, they used feature selection algorithms to reduce the number of features in the machine learning classifier. The computational complexity of this approach is low. However, false positive rate is high in this approach.

According to Gheorghe et al. [112], the consumption of system resources in some malware apps are minimal. Hence the system metric values (CPU usage, memory usage) are not high for these applications. In such cases, a system metric-based malware detection mechanism may fail in detecting these kinds of malware.

4.2.2   Network packet analysis (Category 2)

In these mechanisms, the network packets are analyzed for finding malicious behavior. These approaches are very accurate in detecting the malicious applications which try to communicate with the remote servers in the background.

In [73], the network packet level information such as average packet size, ratio of incoming to outgoing bytes etc. are used for predicting the malicious behavior. In this approach, a trained ML classifier is used to predict whether the application is malicious or not based on the network traffic features. The cost of feature extraction is very high in this approach.

In AppFA[119], He et al. proposed a mechanism to construct the network behavioral profiles of an application using a clustering algorithm. Then, these network behavioral profiles are compared with historical data and profiles of selected peer groups (apps having similarity in size, category etc.) for identifying the malicious behavior. Malware detection rate (0.734) is low in this approach.

Some malware do not need the Internet connection for performing their activities. For example, some malware send SMS to premium numbers in the background. These kinds of malware cannot be detected by analyzing network packets. This is a major drawback of this approach.

4.2.3   Sensitive API call analysis (Category 3)

It is required for an application to invoke sensitive API calls for performing privileged operations in a device. Hence, the malicious behavior of an application can be identified from the API calls invoked by the application during its runtime.

In [82], Bao et al. suggested a mechanism to detect Android malware by analyzing the sensitive API calls. In this approach, the malicious behavior of an unknown application is identified by checking the presence of malicious API calls. This approach cannot detect evolving malware apps.

In [130], Kim et al. proposed a mechanism to detect Android malware apps based on a suffix tree which contains API call subtraces. In this approach, Hidden Markov Model is used for generating the probabilistic confidence values from the suffix tree for identifying the malicious behavior. This approach cannot detect malware apps which contain privilege escalation exploits.

This mechanism is very accurate in detecting the malware apps which try to make a communication with the sensitive resources in a device by generating a sensitive API call sequence. Sometimes benign applications (goodware) invoke sensitive API calls for performing privileged operations. In those cases, benign applications may get wrongly classified as malware.

4.2.4 System call analysis (Category 4)

In this section, we discuss about the system call-based dynamic analysis mechanisms for detecting malware applications. Malware apps do not require user triggers for invoking sensitive APIs [101]. The invocation of API calls without user triggers get reflected in the corresponding system call sequence [211]. Hence, system call analysis is a powerful approach for dynamic malware detection. The system call-based dynamic analysis mechanisms can be classified into:

- System Call Frequency or TF-IDF (Term Frequency-Inverse Document Frequency) Based Methods;

- System Call Dependency Graph or Markov Chain Based Methods;

- System Call Phylogeny Based Methods;

- System Call Behavioral or Sequence Analysis Based Methods .

4.2.4.1 System call frequency or TF-IDF-based methods

In these mechanisms, the counts of system calls or TF-IDF values are analyzed using machine learning algorithms for finding the malicious behavior. It is computationally easy to construct a feature vector based on the counts or TF-IDF values of system calls in the system call sequence of an application.

Crowdroid [89] is a behavior based dynamic malware detection mechanism based on a cloud architecture. In this approach, a client application collects all the system call events from a device and sends to a remote server. Then the server preprocesses this system call data and uses k-means clustering algorithm to determine whether the application is malicious or not. This approach can be defeated by employing system call injection attack. That is, an attacker can inject irrelevant system calls such as information maintenance system calls, erroneous system calls, etc. in the system call sequence of an application.

Amamra et al. [71] proposed a mechanism for detecting Android malware applications by analyzing the frequencies of system calls which influence the behavior of an application. In this approach, a binary machine learning classifier is trained with the frequencies of behavioral system calls produced by known malware and goodware applications. The classifier is used to predict whether an application is malware or not based on the frequencies of behavioral system calls. The authors used a limited dataset (100 malware and 100 goodware) to evaluate their approach.

Vinod et al. [199] suggested novel feature selection methods for finding the most relevant system calls which provide more accuracy in detecting malware and goodware apps. In this approach, a machine learning classifier is used to predict whether an application is malware or not based on the TF-IDF (Term Frequency- Inverse Document Frequency) values of the selected system calls. In this approach, the system call relationship information is not considered.

In [209], Xiao et al. proposed a mechanism to detect Android malware applications by analyzing system calls during its runtime. A pre-trained machine learning classifier is used to predict the malicious behavior of an unknown application based on the system call co-occurrence values and the frequency values of each system calls in a sequence. In this approach, the dimensionality of feature vector is high.

Canfora et al. [91] proposed a mechanism to detect Android malware application by analyzing the counts of the fixed length contiguous subsequences in the system call sequence of an application. They used a trained SVM classifier for identifying whether an application is a malware or not based on the counts of fixed length contiguous system call subsequences. This approach cannot detect unknown malware families.

4.2.4.2   System call dependency graph or markov chain-based methods

In these mechanisms, system call graphs or Markov chain sequences of system calls are analyzed for finding the malicious behavior. Here, system call dependency information is considered for finding malicious behavior.

Maline [100] is a tool for detecting malware applications in Android devices based on system call sequence. In Maline, a trained classifier is used to predict whether an application is malicious or not by finding the dependencies of each system calls and their frequencies of occurrence. This approach can be defeated by employing system call injection attack. That is, an attacker can inject irrelevant system calls such as information maintenance system calls, erroneous system calls, etc. in the system call sequence of an application and defeat the detection mechanism.

In [122], Hu et al. proposed a mechanism to detect Android malware applications from system call graphs. In this approach, they used a deep learning classifier for identifying whether an application is malicious or not based on the system call graph-based features. This mechanism rely on deep learning algorithms for malware detection. It requires a very large set of training data.

Xiao et al. [208] suggested a mechanism for detecting Android malwares from the Markov chain sequence of system calls of the application. A back propagation neural network [120] is trained with the system call state transition probabilities using several goodware and malware applications. This neural network [120] is used to predict the malicious behavior of the application. In this approach, the dimensionality of feature vector is high.

In CSCdroid [219], the Markov chain system call sequence of an application is analyzed for detecting malware applications. This approach has two steps. In the first step, they constructed a system call Markov chain sequence by replacing all normal system calls in a sequence with a specific one. In the second step, a binary machine learning classifier such as SVM is used for detecting the malicious behavior of an application from its system call Markov chain state transition probability matrix. This approach can be defeated by employing system call injection attack. That is, an attacker can inject irrelevant system calls such as information maintenance system calls, erroneous system calls, etc. in the system call sequence of an application and defeat the detection mechanism.

Bandari et al. [85] proposed a mechanism to detect Android malware application from its Markov chain system call sequence. In this approach, they extracted semantically relevant paths from system call Markov chain graph and used it as a feature vector of a machine learning classifier for finding the malicious behavior. It requires at least 30 minutes to extract the semantically relevant paths from system call Markov chain graph of an application.

4.2.4.3 System call phylogeny-based methods

In these mechanisms, the system call phylogeny relationship among malware samples are analyzed for finding malicious behavior.

Bernadi et al. [83] proposed a mechanism to detect Android malware apps from System Call Execution Fingerprint (SEF)-based feature vectors. This SEF (System Call Execution Fingerprint)-based features can be used to characterize the malicious behavior and trace out the malware phylogeny. In this mechanism, a trained machine learning classifier is used to identify the malicious behavior of an application from the SEF-based features. It is practically very difficult to collect multiple system call log of a single application and construct the feature vector.

4.2.4.4 System call behavior or sequence analysis-based methods

In these mechanisms, the system call sequence or subsequences are analyzed for finding malicious behavior.

Xiao et al. [210] proposed an LSTM (Long Short Term Memory)-based mechanism for detecting Android malware applications from its system call sequences. In this approach, two trained LSTM classifiers corresponding to malware and goodware system call sequences are used to predict the malicious behavior from the system call sequence generated by an application. It is possible for an attacker to reorder the system call sequence by reordering the source code. Further, it requires a very large training dataset.

In [99], Dash et al. suggested a mechanism called DroidScribe which makes use of Copperdroid [194] to reconstruct high level behavior from the system calls and these behaviors are used as features of a machine learning classifier for identifying the malicious applications. It is required to customize emulators for reconstructing the malicious behavior from system call sequence.

All the existng system call analysis mechanisms can be defeated by employing system call hijacking attack [168]. That is, it is possible for an attacker to defeat system call-based malware detection mechanisms by changing system call names in the system call table. In such cases, a malware app may get wrongly flagged off as a goodware and vice versa.

In this section, we discussed various dynamic analysis mechanisms and their limitations. Dynamic analysis mechanisms can detect a malicious application only when the application expresses malicious behavior at least once in the analysis time. Hence, it is possible for a malware developer to defeat dynamic analysis mechanism by employing the update attack after the analysis time.

4.3   Hybrid Analysis

Hybrid analysis is a combination of static and dynamic analysis. The comparative analysis of existing hybrid analysis mechanisms are given in Table 4.2. The existing hybrid detection mechanisms can be classified into two categories. They are:

• Hybrid Detection Based on a Single Classifier (Category 1);

• Hybrid Detection Based on Ensemble of Classifiers (Category 2).

TABLE 4.2   Comparative analysis of hybrid malware detection mechanisms.

Category	Performance	Limitations
Category 1 [160]	F1 Score: 0.920	Prone to API call injection attacks
Category 1 [217]	Accuracy: 0.967	Requires a very large training dataset
Category 1 [74]	Accuracy: 0.948	System calls are not considered
Category 1 [214]	Accuracy: 0.947	Requires a very large training dataset
Category 1 [93]	Accuracy: 0.938	High dimensionality of feature vector
Category 2 [143]	Accuracy: 0.897	High dimensionality of feature vector
Category 2 [76]	Static Accuracy : 0.980 Dynamic Accuracy : 0.820 Hybrid Accuracy : N/A	Low accuracy in dynamic analysis

4.3.1   Hybrid detection based on a single classifier (Category 1)

In these mechanisms, both static and dynamic features are used as features of a machine learning classifier for identifying the malicious behavior. In these mechanisms, the feature fusion is carried out by the classifier itself. Hence, feature fusion mechanisms are not required in this approach.

In [74], Arora et al. suggested a mechanism to detect Android malware from the permissions and the network traffic-based features. In this approach, the permissions and network traffic features are used in the frequent pattern growth algorithm to detect the malicious behavior. This approach can only detect the malware apps which try to communicate with a remote server.

In [160], Onwuzurike et al. suggested a hybrid analysis mechanism for detecting Android malware apps. In this approach, abstracted API call Markov chain sequences from static and dynamic analysis are combined for detecting the malicious behavior. It is possible for an attacker to inject additional API calls in the system call sequence. In such cases, a malware app may get falsely flagged off as a goodware.

In StormDroid [93], static features such as permissions, sensitive API calls and their sequences and dynamic features from the droidbox such as sendSMS, recvnet, sendnet, etc. are used in a machine learning classifier for identifying the malicious behavior. In this mechanism, the high dimensionality of feature vector unnecessarily increases the storage space and processing time of the classifier.

In DroidDetector [217], static features such as sensitive APIs, requested permissions and dynamic features obtained from the droidbox are used in a deep learning model for identifying the malicious behavior. This mechanism rely on deep learning algorithms for malware detection. It requires a very large set of training data.

Xu et al. suggested a malware detection mechanism called HADM [214]. In this approach, hybrid features such as suspicious calls, intent filters, network APIs, APIs related to advertising networks such as Google ads and AdMob, instruction sequences and system calls are used as feature vectors of a DNN (Deep Neural Network) classifier for identifying the malicious behavior. It requires to train DNN classifier on a large set of high dimensional feature vectors. Due to the high complexity, it is infeasible to use this mechanism to use in real time malware detection.

In these mechanisms, there are too many features used in a single classifier. Hence, it is possible for a malware developer to defeat these mechanisms by injecting the features which are frequently found in goodware apps.

4.3.2 Hybrid detection based on ensemble classifiers (Category 2)

In these mechanisms, ensembles of classifiers are used for identifying malicious behavior from static and dynamic features. These mechanisms are highly accurate than others.

Samadroid [76] is a hybrid malware detection model for Android devices. Samadroid works in two steps. In the first step, the static features such as requested hardware components, requested permissions, used permissions, app components, intent filters and API calls are extracted from the source code and the dynamic feature such as file and network related system calls such as open, read, etc. produced by the application are collected after executing it. In the second phase, these static and dynamic features are preprocessed and used as inputs to two machine learning classifiers such as SVM for identifying whether the application is a malware or not. An application is treated as a malware if it is flagged off as malicious by both static and dynamic classifiers. The accuracy of this approach is low due to the low accuracy in its dynamic analysis part.

In Omnidroid [143], voting based ensembles of machine learning classifiers are used for fusing the static and dynamic features extracted from malware and goodware applications. In this approach, the dimensionality of feature vector is high. The high dimensionality of feature vector unnecessarily increases the storage space and processing time of the classifier.

There are multiple classifiers used in these mechanisms. Hence, it is required to develop innovative mechanisms to combine the prediction results of distinct machine learning classifiers for accurate malware detection.

4.4   Correlation Among Static and Dynamic Features

Android malware applications have a malicious code along with appropriate permissions for accessing sensitive resources from the device without the proper permission of a user. According to Zhang et al. [218], the source code level API calls can determine the underlying semantics of an application. These APIs are protected by some permissions which need to be declared in the manifest file [106]. It is possible for a developer to declare permissions without API calls and vice versa. Hence, using either permissions or API calls alone as features is not enough to detect the malicious behavior of an application. Therefore, it is required to combine both API calls and permissions for accurately detecting malware applications. Malware applications do not require user triggers for invoking sensitive API calls unlike goodware [101, 211, 176]. This automated invocation of API calls gets reflected in a system call sequence [211]. It is known that, an application generates system calls in accordance with the execution of API calls during its runtime [183]. Hence, we can conclude that static features such as API calls, permissions and dynamic features such as system calls are the relevant features for detecting malicious applications.

Whenever an application process invokes an API call, the system process will check the corresponding permissions required for it [106]. Then, it will allow that API to execute if the permission is granted. For instance, the system process allows sendTextMessage() API call to execute only if SEND_SMS permission is granted. Hence, we can assume that an API call is conditionally dependent on its permission. Hardware Abstraction Layer (HAL) accepts the invoked API calls and communicates with the Linux kernel by generating system calls [86]. Hence, we can assume that there exists some conditional dependencies between system calls and API calls. The conditionally dependent execution of API call/s is given in Figure 4.1.

FIGURE 4.1   Conditional dependencies among features.

4.4.1   Tree augmented Naive Bayes (TAN) model

The naïve Bayes classifier builds a simple probabilistic model based on the assumption that the features are independent of each other. In reality this assumption may not hold. If we can incorporate the information of dependencies between the features within the naïve Bayes model, the classification accuracy can be improved. Tree augmented naïve Bayes (TAN) is one such model where the features are not independent as in the naive Bayes. In TAN, each feature conditionally depends on the class and (one) another feature from the feature set.

In TAN model the basic structure of the naive Bayes model is retained. However, the class node and all the feature nodes are connected by directed edges. Thus, it will take into consideration all the features while calculating the conditional probability $Pr(C|A_1 \ldots A_n)$, where C is the class variable and $A_1, \ldots, A_n$ are the features. Except for the root variable (the class), each variable in the tree will have one or two parents: one is class node and the other one, when present, is another random variable corresponding to a feature. As interaction between the attributes have been limited to only one, the computational complexity of this model is low.

To construct the tree structure we need to find the features which are correlated. This is how the parent of each feature is found out. In this way, the problem of constructing a maximum likelihood tree can be reduced to the construction of a maximum weighted spanning tree in a graph. In order to find the most correlated features mutual information is calculated between each pair of attributes which form the weight of the edges. The edges are added between the features which are highly interdependent. If there are n features, there will n nodes and $n - 1$ edges in the graph. The resultant graph with mutual information on the edges will form a maximum weight spanning tree. For constructing a TAN model, we need to measure the degree of correlation between the predictors (random variables corresponding to features) with the help of a previously known dataset with class labels $\mathcal{E}$. This is done by measuring the conditional mutual information among all pair of predictors in that labelled dataset. The conditional mutual information $I(A_i, A_j|E)$ between two random variables A_i, and A_j is computed as,

$$
\begin{aligned}
I(A_i, A_j|E) = \sum_{x,y,k} (Pr(A_i = x, A_j = y, E = k). \\
\log \frac{Pr(A_i = x, A_j = y|E = k)}{Pr(A_i = x|E = k)Pr(A_j = y|E = k)}),
\end{aligned}
\tag{4.1}
$$

where the probability values $Pr(A_i = x, A_j = y, E = k)$, $Pr(A_i = x, A_j = y|E = k)$ and $Pr(A_i = x|E = k)Pr(A_j = y|E = k)$ are obtained from the dataset. After calculating the CMI values, the conditional dependencies among the predictor variables are modeled as a tree. This tree can be used for the classification

4.5 Hybrid Analysis with TAN Classifier

In the previous section, we saw that there exists conditional dependencies among the features, API calls, permissions and system calls. However, using conditionally dependent static and dynamic features as a feature vector in a machine learning classifier for hybrid analysis can lead to multicollinearity problem[70] . In [188], the mean of output probability values of three naïve Bayes classifiers corresponding to API calls, permissions and system calls is used for malware detection. However, in this approach, outlier probability values may affect the performance of the model. In [191],

TAN classifier is used to fuse the prediction results of the classifiers corresponding to API calls, permissions and system calls. In this section, we explain about this TAN classifier

In order to overcome these limitations, in this chapter, we discuss a TAN model[191] to combine the classifier output variables corresponding to the static features such as API calls, permissions and the dynamic features such as system calls based on their conditional dependencies for predicting the malicious behavior. Hence we can use a Tree Augmented Naive Bayes (TAN) model to combine the probabilities of these correlated features to identify whether it is a malware or not. TAN is a restricted Bayesian network based on the Bayes theorem. TAN can be used for modeling the conditional dependency relationships among random variables as a tree. Further, it has superior classification performance than naive Bayes classifier. This TAN-based model can capture the interdependencies between static and dynamic features for predicting the malicious behavior. Suppose, in an application, sendTextMessage() API call is declared without SEND_SMS permission. In this situation, sendTextMessage() API call will not execute. Hence, without the execution of sendTextMesssage() API call, the malicious system calls will not generate. In such cases, only the API call feature vector will have non zero values. Hence the API call-based classifier may classify this application as anomalous, whereas the permission based classifier and the system call based classifier may classify this application as goodware. Hence the TAN model which combines the outputs of these three classifiers will correctly classify the application as goodware. However, in the integrated feature fusion approaches, the classifiers may wrongly classify the app as malware because of the presence of API calls in the feature vector. Ridge regularized logistic regression models (RRLR) are used for training and testing in this TAN model. The advantage of the RRLR classifier is its ability to predict anomalous behavior even in the presence of noisy samples in the dataset [172]. In RRLR classifiers , regularization is used to tune the parameters for minimizing the prediction errors caused by the noisy applications.

We will now discuss the details of the TAN based hybrid malware detection mechanism [191]. This mechanism has two phases. In the first phase, we will estimate the probabilities of anomalous behaviors in static API calls, permissions and the system calls using RRLR (Ridge Regularized Logistic Regression) classifiers. In the second phase, we combine these probabilities using TAN for effectively predicting the malicious behavior. The hybrid malware detection mechanism is given in Figure 4.2.

4.5.1 Dependencies among API calls, permission and system calls

In this section, we illustrate conditional dependencies among these API calls, permissions and system calls with the help of a dataset.

4.5.2 Ridge regularized logistic regression (RRLR)

Logistic regression is the fitting of a logit function or an s-curve logistic to a dataset in order to calculate the probability of the occurrence of a categorical event based

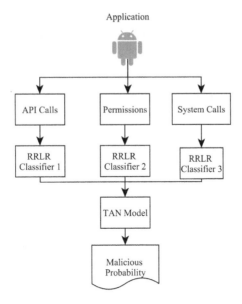

FIGURE 4.2 Hybrid malware detection mechanism.

on the values of a set of independent variables. A logistic model might predict the likelihood of a given application to be a malware as a function of its static or dynamic features.

A logistic regression in one independent variable has the general form,

$$h_\delta(x) = \frac{1}{1 + e^{-(\delta_0 + \delta_1 x)}}.$$

Logistic regression can be generalized to multiple independent variables as,

$$h_\delta(X) = \frac{1}{1 + e^{-(\delta_0 + \delta_1 x_1 + \cdots + \delta_n x_n)}} = \frac{1}{1 + e^{-\delta^T X}},$$

where $\delta = (\delta_1, \ldots, \delta_n)^T$ and $X = (x_1, \ldots, x_n)^T$. The probability of occurrence of a dependent variable Y given X can be directly estimated as,

$$h_\delta(X) = Pr(Y|X) = \frac{1}{1 + e^{-\delta^T X}},$$

where $\delta = (\delta_1, \delta_2, \ldots, \delta_n)^T$ are the regression parameters. These regression parameters are estimated from the training data.

Ridge and Lasso are two regularization techniques to reduce or shrink the coefficients in the resulting regression. This reduces the variance in the model and can avoid over fitting problem. Over fitting results in a model failing to generalize. That is the model works well on a data used to train the model, whereas it fails to perform on a new set of data. The logistic regression coefficients $\delta_0, \ldots, \delta_n$ are found by minimising the negative log likelihood. Ridge and lasso regularization work by adding a penalty term to the log likelihood function. In the case of ridge regression, the penalty term is $\sum \delta_i^2$ and in the case of lasso, it is $\sum |\delta_i|$. In this chapter, we will employ the ridge regularization to estimate the probabilities.

Let $D = \{(X_i, Y_i) : i = 1, 2, \ldots, m\}$ be a labeled dataset, where each X_i is an n dimensional vector and Y_i denotes its label. The regression parameter vector $\delta = (\delta_1, \delta_2, \ldots, \delta_n)$ can be estimated from the dataset D through ridge regularization. The parameter δ is estimated as,

$$\arg\max_{\delta} \sum_{i=1}^{m} \log(Pr(Y_i|X_i : \delta)) - \gamma \sum_{i=1}^{n} \delta_i^2,$$

where γ is a penalty value calculated by cross validation approach [172]. $Pr(Y_i|X_i : \delta))$ is computed as,

$$Pr(Y_i|X_i : \delta) = h_{\delta}(X_i)^{Y_i}(1 - h_{\delta}(X_i))^{(1-Y_i)}.$$

4.5.3 Probability estimation

RRLR classifier is used to find the probability of anomaly (malicious behavior) associated with a q-dimensional feature vector X (based on permissions or API calls or system calls). Let $Y = 1$ represents anomalous (malicious) and $Y = 0$ represents non anomalous (non malicious). Then, $Pr(Y = 1|X)$ is computed as:

$$Pr(Y = 1|X) = \frac{1}{1 + exp(-\delta^T X)},$$

where $\delta = (\delta_1, \delta_2, \ldots, \delta_q)$ are the regression parameters. These regression parameters are estimated during the training phase.

Let $D = \{(X_i, Y_i) : i = 1, 2, \ldots, m\}$ be a labeled dataset, where $X_i = (X_{i1}, \ldots, X_{iq})$ be the q dimensional feature vector correspond to the ith element and $Y_i \in \{0, 1\}$ denotes its label (1 represents anomalous and 0 represents non anomalous). During the training time, we estimate regression parameters $\delta = (\delta_1, \delta_2, \ldots, \delta_q)$ from the dataset D as $\arg\max_{\delta} \sum_{i=1}^{m} log(Pr(Y_i|X_i : \delta)) - \gamma \sum_{i=1}^{q} |\delta_i^2|$, where γ is a penalty value calculated by cross validation approach [172].

4.5.4 Anomaly detection

In this phase, we will inspect API calls, requested permissions and system calls generated by the application for finding the anomalous behavior. In TAN-based model, anomaly in API, permission and system call based features refer to their abnormal or suspicious behavior. The anomalous API calls, permissions and system calls are invoked for performing high privileged operations in a device. The source code of a malware application may contain risky API calls (anomalous) to perform high privileged operations such as calling phone, sending SMS, taking pictures, recording audio and so on. In order to perform these highly privileged operations, an application requires appropriate risky permissions (anomalous) such as CALL_PHONE and SEND_SMS which need to be declared in the manifest file. These kinds of risky permissions are required for the execution of API calls (anomalous) in the runtime. In order for the execution of these risky API calls (anomalous) in the runtime, infrequent risky system call sequences (anomalous) will be generated. These system call sequences (anomalous) are intended to perform the high privileged operations in the device. Here, we train three distinct ridge regularized LR classifiers corresponding to API calls, permissions and system call sequences of a set of malware and goodware applications.

4.5.4.1 App permission analysis

In this phase, we will analyze the permissions declared in the app manifest file for estimating the abnormal behavior. The process consists of two steps. In the first step, we will train an RRLR (Ridge Regularized Logistic Regression) classifier [138] with the permissions of known malware and goodware applications. In the second step, the trained RRLR classifier is used for estimating the malicious probability of unknown application based on the permission based features.

Let $\mathcal{A}=(\mathcal{A}_1, \ldots, \mathcal{A}_u)$ represents the feature vector based on the permissions mentioned in the app manifest file and $\mathcal{E} \in \{0, 1\}$ denotes the binary random variable of an application $\mathcal{F}$ in which 1 represents anomalous and 0 represents non anomalous. $\mathcal{A}_i$ in $\mathcal{A}$ is computed as:

$$\mathcal{A}_i = \begin{cases} 1, & \text{if } i^{th} \text{ permission is present in } \mathcal{F}; \\ 0, & \text{otherwise.} \end{cases}$$

Let $\mathcal{T}_1$ be a threshold value lying between 0 to 1. Let $\mathcal{Z}_1$ be a binary value indicating whether the app manifest permissions of an application is considered as anomalous or not by the RRLR classifier $\mathcal{L}_1$. That is,

$$\mathcal{Z}_1 = \begin{cases} 1 & \text{if } Pr(\mathcal{E} = 1|\mathcal{A}) \geq \mathcal{T}_1; \\ 0 & \text{otherwise.} \end{cases}$$

4.5.4.2 Static API function call analysis

In this phase, we will analyze the static API function calls of an application to predict the abnormal behavior. The process consists of two steps. In the first step, we will train an RRLR (Ridge Regularized Logistic Regression) classifier [138] with the static API function calls of known malware and goodware applications. In the second step, the trained RRLR classifier is used for estimating the malicious probability of unknown application based on the static API function calls in it.

Assume that $\mathcal{B}=(\mathcal{B}_1, \ldots, \mathcal{B}_v)$ denotes the features based on static API function calls and $\mathcal{E} \in \{0, 1\}$ denotes the binary random variable of an application $\mathcal{F}$ in which 1 represents anomalous and 0 represents non anomalous. $\mathcal{B}_i$ in $\mathcal{B}$ is computed as:

$$\mathcal{B}_i = \begin{cases} 1, & \text{if } i^{th} \text{ API call is present in } \mathcal{F}; \\ 0, & \text{otherwise.} \end{cases}$$

Let $\mathcal{T}_2$ be a threshold value lying between 0 to 1. Let $\mathcal{Z}_2$ be a binary value indicating whether the static API function calls of an application is considered as anomalous or not by the RRLR classifier $\mathcal{L}_2$. That is,

$$
\mathcal{Z}_2 = \begin{cases} 1 & \text{if } Pr(\mathcal{E} = 1|\mathcal{B}) \geq \mathcal{T}_2; \\ 0 & \text{otherwise.} \end{cases}
$$

4.5.4.3   System call analysis

In this phase, we will analyse the counts of individual system calls in a system call sequence. The process consists of two steps. In the first step, we will train an RRLR (Ridge Regularized Logistic Regression) classifier [138] with the system call counts of known malware and goodware applications. In the second step, the trained RRLR classifier is used for estimating the malicious probability of unknown application based on the counts of system calls in it.

Assume that $C=(C_1,\ldots,C_w)$ denotes the system call count based feature vector and $\mathcal{E} \in \{0,1\}$ denotes the binary random variable of an application $\mathcal{F}$ in which 1 represents anomalous and 0 represents non anomalous.

Let $\mathcal{T}_3$ be a threshold value lying between 0 to 1. Let $\mathcal{Z}_3$ be a binary value indicating whether the system call counts of an application is considered as abnormal or not by the RRLR classifier $\mathcal{L}_3$. That is,

$$
\mathcal{Z}_3 = \begin{cases} 1 & \text{if } Pr(\mathcal{E} = 1|C) \geq \mathcal{T}_3; \\ 0 & \text{otherwise.} \end{cases}
$$

4.5.5   Malware detection using TAN-based model

In this section, we combine the outputs of machine learning classifiers using Tree Augmented Naive Bayes algorithm [97] [111]. Assume that $\mathcal{F}$ is an unknown application and $\mathcal{A}, \mathcal{B}, C$ correspond to the manifest permission based feature vector, static API call based feature vector and system call count based feature vector respectively. Assume that $\mathcal{Z}_i$ for $i = 1, 2, 3$ denote the binary valued random variables indicating whether the application $\mathcal{F}$ has been declared as anomalous or not by the RRLR classifiers $\mathcal{L}_i$, for $i = 1, 2, 3$ respectively. The TAN model is given in Figure 4.3. The Tree Augmented naive Bayes algorithm [97] is given below:

1. Measure the Conditional Mutual Information, $I(\mathcal{Z}_i, \mathcal{Z}_j|\mathcal{E})$ between $\mathcal{Z}_i$ and $\mathcal{Z}_j$ for $i \neq j$. The conditional mutual information $I(\mathcal{Z}_i, \mathcal{Z}_j|\mathcal{E})$ between two ML classifier outputs $\mathcal{Z}_i$ and $\mathcal{Z}_j$ is calculated as,

$$
I(\mathcal{Z}_i, \mathcal{Z}_j|\mathcal{E}) = \sum_{u=0}^{1}\sum_{v=0}^{1}\sum_{k=0}^{1}(Pr(\mathcal{Z}_i = u, \mathcal{Z}_j = v, \mathcal{E} = k).
$$
$$
\log \frac{Pr(\mathcal{Z}_i = u, \mathcal{Z}_j = v|\mathcal{E} = k)}{Pr(\mathcal{Z}_i = u|\mathcal{E} = k)Pr(\mathcal{Z}_j = v|\mathcal{E} = k)}). \tag{4.2}
$$

2. Construct an undirected graph with $\mathcal{Z}_1, \mathcal{Z}_2, \mathcal{Z}_3$ as the vertices connected by edges with weights $I(\mathcal{Z}_i, \mathcal{Z}_j|\mathcal{E})$.

3. Extract an undirected maximum spanning tree from the obtained undirected graph.

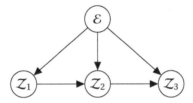

FIGURE 4.3 TAN Model for malicious probability estimation.

4. Transform the undirected maximum spanning tree to directed maximum spanning tree by arbitrarily choosing a node as root and setting directed edges outward from it.

5. Construct a TAN model by adding a vertex $\mathcal{E}$ and adding edges from $\mathcal{E}$ to all remaining vertices.

The malicious probability value $Pr(\mathcal{E} = 1|\mathcal{Z}_1, \mathcal{Z}_2, \mathcal{Z}_3)$ of an application is computed using TAN as follows,

$$Pr(\mathcal{E} = 1|\mathcal{Z}_1, \mathcal{Z}_2, \mathcal{Z}_3) = \frac{Pr(\mathcal{E} = 1).Pr(\mathcal{Z}1, \mathcal{Z}_2, \mathcal{Z}_3|\mathcal{E} = 1)}{Pr(\mathcal{Z}_1, \mathcal{Z}_2, \mathcal{Z}_3)}. \tag{4.3}$$

The probability $Pr(\mathcal{Z}_1, \mathcal{Z}_2, \mathcal{Z}_3|\mathcal{E} = 1)$ is calculated as,

$$Pr(\mathcal{Z}_1, \mathcal{Z}_2, \mathcal{Z}_3|\mathcal{E} = 1) = Pr(\mathcal{Z}_1|\mathcal{E} = 1).Pr(\mathcal{Z}_2|\mathcal{Z}_1, \mathcal{E} = 1).Pr(\mathcal{Z}_3|\mathcal{Z}_2, \mathcal{E} = 1).$$

The probability $Pr(\mathcal{Z}_1, \mathcal{Z}_2, \mathcal{Z}_3)$ is calculated as,

$$\begin{aligned} Pr(\mathcal{Z}_1, \mathcal{Z}_2, \mathcal{Z}_3) &= Pr(\mathcal{E} = 1).Pr(\mathcal{Z}_1, \mathcal{Z}_2, \mathcal{Z}_3|\mathcal{E} = 1) \\ &+ Pr(\mathcal{E} = 0).Pr(\mathcal{Z}_1, \mathcal{Z}_2, \mathcal{Z}_3|\mathcal{E} = 0). \end{aligned}$$

Let $\mathcal{T}$ be a threshold value lying between 0 and 1. If the malicious probability value $Pr(\mathcal{E} = 1|\mathcal{Z}_1, \mathcal{Z}_2, \mathcal{Z}_3)$ exceeds $\mathcal{T}$, then the application is declared as a malware.

4.6 Experiments and Analysis

In this section, we discuss about the performance of the TAN-based hybrid model in the existing malware and goodware datasets. We have taken 1000 malware applications from Drebin [75] and AMD [203] and 1000 goodware applications from Google Play (GP) [28] for evaluating the performance of this approach. The implementation of this approach was carried out in an intel core i5 Windows 10 PC with 8GB memory having a preinstalled Android 4.4 emulator.

In static analysis, API calls and permissions of 3200 apps were extracted. For extracting API calls and permissions, we have used Androguard [98] tool . After

extracting the features, the binary feature vectors were constructed from these extracted API calls and permissions using a python code and saved in separate files.
In the dataset, we observed the presence of some permissions more frequently in
malware applications than in goodware applications and vice versa. We selected
these as our permission based features. The selected permissions are given in
Table 4.3.

TABLE 4.3   Selected permission for malware detection.

ID	Permission	ID	Permission
1	READ_PHONE_STATE	2	WRITE_CONTACTS
3	CALL_PHONE	4	READ_CONTACTS
5	Internet	6	SEND_SMS
7	DISABLE_KEYGUARD	8	PROCESS_OUTGOING_CALLS
9	RECEIVE_BOOT_COMPLETED	10	READ_SMS
11	FACTORY_TEST	12	DEVICE_POWER
13	HARDWARE_TEST	14	CHANGE_WIFI_STATE
15	GET_ACCOUNTS	16	READ_HISTORY_BOOKMARKS
17	WRITE_APN_SETTINGS	18	MODIFY_PHONE_STATE
19	WRITE_HISTORY_BOOKMARKS	20	ACCESS_LOCATION
21	EXPAND_STATUS_BAR	22	WRITE_EXTERNAL_STORAGE
23	RECEIVE_SMS	24	WRITE_SMS
25	ACCESS_WIFI_STATE	26	MODIFY_AUDIO_SETTINGS
27	KILL _BACKGROUND_PROCESS		

We observed the presence of some sensitive API calls frequently in malware applications than in goodware applications and vice versa. We selected these as our API
call-based features. The selected API calls are given in Table 4.4.

For dynamic analysis, we extracted the system call sequences of 2000 apps in our
dataset. The applications in our dataset were installed in an emulator and system call
logs were collected using strace utility. Here, we injected 1000 random events using
monkeyrunner tool while collecting the system call sequence. We used a python code
to transform the system call sequences into system call count-based feature vectors.

After collecting the system call sequences, the system call frequency based feature
vectors were constructed and saved in a file. Wang et al. [202], showed that memory
management calls, error system calls and information maintenance system calls do
not hold any significant impact on the malicious behavior. So, these system calls can
be avoided from the recorded system calls for effective identification of anomalous
behavior. The selected system calls and their alternate notations are given in Table
4.5.

4.6.1   Training phase

We randomly selected 500 malware and 500 goodware samples for training the three
ridge regularized LR classifiers $\mathcal{L}_1$, $\mathcal{L}_2$, $\mathcal{L}_3$. The different steps in training process
are given as follows:

TABLE 4.4 Selected API calls for malware detection.

ID	API Call	ID	API Call
1	getNetworkType	2	getNetworkOperator
3	loadClass	4	getClassLoader
5	getMethod	6	getLongitude
7	getLatitude	8	createFromPdu
9	getInputStream	10	getOutputStream
11	getWifiState	12	abortBroadCast
13	getAccountName	14	RequestFocus
15	getSubscriberId	16	getDisplayOriginatingAddress
17	sendTextMessage	18	getCredential
19	getDisplayMessageBody	20	getPackageInfo
21	getIMEI	22	getLastKnownLocation
23	getAppPackageName	24	takepicture
25	getCookies	26	killProcess
27	exec	28	getMessage

TABLE 4.5 Selected system calls for malware detection.

ID	Notation	System Call	ID	Notation	System Call
1	A	recvfrom	2	B	write
3	C	ioctl	4	D	read
5	E	sendto	6	F	dup
7	G	writev	8	H	pread
9	I	close	10	J	socket
11	K	bind	12	L	connect
13	M	mkdir	14	N	access
15	O	chmod	16	P	open
17	Q	rename	18	R	fchown32
19	S	unlink	20	T	pwrite
21	U	unmask	22	V	lseek
23	W	fcntl	24	X	recvmsg
25	Y	sendmsg	26	Z	epoll
27	A1	dup2	28	A2	fchown
29	A3	readv	30	A4	chdir
31	A4	execve			

- Estimation of threshold for $\mathcal{L}_1, \mathcal{L}_2, \mathcal{L}_3$;

- Conditional probability estimation.

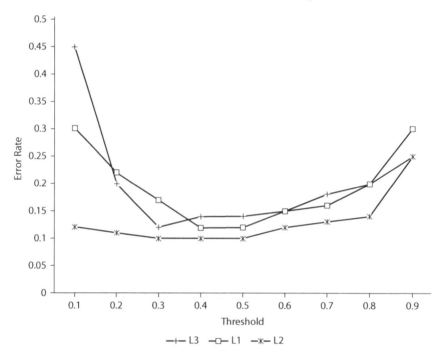

FIGURE 4.4   Classification error rate against different threshold values.

4.6.1.1   Estimation of threshold for $\mathcal{L}_1, \mathcal{L}_2, \mathcal{L}_3$

We split our training dataset into two subsets in which one of them is used to train the classifiers and other is used for validation. We used 250 malware and 250 good-ware samples for training the classifiers $\mathcal{L}_1$, $\mathcal{L}_2$ and $\mathcal{L}_3$. Then, we tested them with the remaining malware and goodware samples and classifier output probabilities $Pr(\mathcal{E} = 1|\mathcal{A})$, $Pr(\mathcal{E} = 1|\mathcal{B})$ and $Pr(\mathcal{E} = 1|C)$ were recorded. The error rate (discussed in Chapter 2) of API, permission and system call based classifiers against different threshold values (probability) are given in Figure 4.4. From Figure 4.4, we can see that, the classifiers can effectively detect anomalies in any threshold (probability) value between 0.3 and 0.6. Here, we chose 0.5 as the values of the thresholds $\mathcal{T}_1$, $\mathcal{T}_2$ and $\mathcal{T}_3$ for the classifiers $\mathcal{L}_1$, $\mathcal{L}_2$ and $\mathcal{L}_3$ respectively .

4.6.1.2   Conditional probability estimation

In this phase, we determine the conditional probabilities among the classifier outputs $\mathcal{Z}_1$, $\mathcal{Z}_2$ and $\mathcal{Z}_3$ using the TAN model. For this, we used the validation set of 250 malware apps and 250 goodware apps. The static (API calls, permissions) and dynamic features (system calls) of applications in the validation dataset are given as inputs to the classifiers $\mathcal{L}_1$, $\mathcal{L}_2$ $\mathcal{L}_3$ and the classifier outputs $\mathcal{Z}_1$, $\mathcal{Z}_2$, $\mathcal{Z}_3$ were recorded. Then, we used TAN model for computing the conditional probability values among $\mathcal{Z}_1$, $\mathcal{Z}_2$

and $\mathcal{Z}_3$. Initially, we calculate conditional mutual information $I(\mathcal{Z}_i, \mathcal{Z}_j|\mathcal{E})$ between $\mathcal{Z}_i$ and $\mathcal{Z}_j$ for $i \neq j$. Then, we construct an undirected graph with $\mathcal{Z}_1$, $\mathcal{Z}_2$ and $\mathcal{Z}_3$ as the set of vertices connected by edges with weights $I(\mathcal{Z}_i, \mathcal{Z}_j|\mathcal{E})$ and extracted a maximum directed spanning tree from it. The undirected graph and maximum spanning directed dependency tree are given in Figures 4.5 and 4.6.

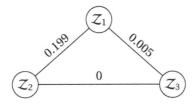

FIGURE 4.5 Undirected dependency graph corresponding to classifier outputs.

FIGURE 4.6 Maximum spanning directed dependency tree corresponding to classifier outputs.

Let $|\mathcal{Z}_3 = j, \mathcal{Z}_2 = i, \mathcal{E} = k|$ be the number of joint occurrences of events $\mathcal{Z}_3 = j$, $\mathcal{Z}_2 = i$ and $\mathcal{E} = k$ and $|\mathcal{Z}_2 = i, \mathcal{E} = k|$ be the number of joint occurrences of events $\mathcal{Z}_2 = i$ and $\mathcal{E} = k$. Then, the probability $Pr(\mathcal{Z}_3 = j, \mathcal{Z}_2 = i|\mathcal{E} = k)$ is calculated as,

$$Pr(\mathcal{Z}_3 = j|\mathcal{Z}_2 = i, \mathcal{E} = k) = \frac{|\mathcal{Z}_3 = j, \mathcal{Z}_2 = i, \mathcal{E} = k|}{|\mathcal{Z}_2 = i, \mathcal{E} = k|}.$$

Let $|\mathcal{Z}_2 = j, \mathcal{Z}_1 = i, \mathcal{E} = k|$ be the number of joint occurrences of events $\mathcal{Z}_2 = j$, $\mathcal{Z}_1 = i$ and $\mathcal{E} = k$ and $|\mathcal{Z}_1 = i, \mathcal{E} = k|$ be the number of joint occurrences of events $\mathcal{Z}_1 = i$ and $\mathcal{E} = k$. Then, the probability $Pr(\mathcal{Z}_2 = j, \mathcal{Z}_1 = i|\mathcal{E} = k)$ is calculated as,

$$Pr(\mathcal{Z}_2 = j|\mathcal{Z}_1 = i, \mathcal{E} = k) = \frac{|\mathcal{Z}_2 = j, \mathcal{Z}_1 = i, \mathcal{E} = k|}{|\mathcal{Z}_1 = i, \mathcal{E} = k|}.$$

Let $|\mathcal{Z}_1 = i, \mathcal{E} = k|$ be the number of joint occurrences of events $\mathcal{Z}_1 = i$ and $\mathcal{E} = k$. Then, the probability $Pr(\mathcal{Z}_1 = i|\mathcal{E} = k)$ is calculated as,

$$Pr(\mathcal{Z}_1 = i|\mathcal{E} = k) = \frac{|\mathcal{Z}_1 = i, \mathcal{E} = k|}{|\mathcal{E} = k|}.$$

Let $|\mathcal{E} = k|$ is the number of occurrences of $\mathcal{E} = k$ and $|\mathcal{E}|$ is the total number of applications in the dataset. Then the probability value $Pr(\mathcal{E} = k)$ is calculated as,

$$Pr(\mathcal{E} = k) = \frac{|\mathcal{E} = k|}{|\mathcal{E}|}.$$

The conditional probabilities are given in Table 4.6, 4.7 and 4.8 respectively.

TABLE 4.6   Conditional probability table for random variable $\mathcal{Z}_1$.

Class Label	$\mathcal{Z}_1$	$Pr(\mathcal{Z}_1 \vert ClassLabel)$
$\mathcal{E} = 1$	1	0.95
$\mathcal{E} = 1$	0	0.05
$\mathcal{E} = 0$	1	0.15
$\mathcal{E} = 0$	0	0.85

TABLE 4.7   Conditional probability table for random variable $\mathcal{Z}_2$.

Class Label	$\mathcal{Z}_1$	$\mathcal{Z}_2$	$Pr(\mathcal{Z}_2 \vert \mathcal{Z}_1, ClassLabel)$
$\mathcal{E} = 1$	1	1	0.89
$\mathcal{E} = 1$	0	1	0.45
$\mathcal{E} = 1$	1	0	0.11
$\mathcal{E} = 1$	0	0	0.55
$\mathcal{E} = 0$	1	1	0.27
$\mathcal{E} = 0$	1	0	0.73
$\mathcal{E} = 0$	0	1	0.20
$\mathcal{E} = 0$	0	0	0.80

TABLE 4.8   Conditional probability table for random variable $\mathcal{Z}_3$.

Class Label	$\mathcal{Z}_2$	$\mathcal{Z}_3$	$Pr(\mathcal{Z}_3 \vert \mathcal{Z}_2, ClassLabel)$
$\mathcal{E} = 1$	1	1	0.85
$\mathcal{E} = 1$	1	0	0.15
$\mathcal{E} = 1$	0	1	0.73
$\mathcal{E} = 1$	0	0	0.27
$\mathcal{E} = 0$	1	1	0.09
$\mathcal{E} = 0$	1	0	0.91
$\mathcal{E} = 0$	0	1	0.06
$\mathcal{E} = 0$	0	0	0.94

    The false positive (FPR) and negative rates (FNR) of the TAN approach against different threshold values (malicious probability) are given in Figure 4.7. From Figure 4.7, we can see that the FPR and FNR values are lowest for any threshold value in the range 0.3 – 0.8. Hence, we can fix any value in the range 0.3 – 0.8 as a threshold. Here, we chose the mid value 0.5 as as the value of the threshold $\mathcal{T}$ for the TAN classifier.

4.6.2   Evaluation phase

The performance of the mechanism was evaluated with 500 malware and 500 goodware applications. The evaluation phase consists of two steps. In the first step, we find out the classifier outputs $\mathcal{Z}_1$, $\mathcal{Z}_2$ and $\mathcal{Z}_3$ from API call, permission and system call-based features of an application. In the second step, we find out the malicious probability value. The performance of this approach is given in Table 4.9. In Table

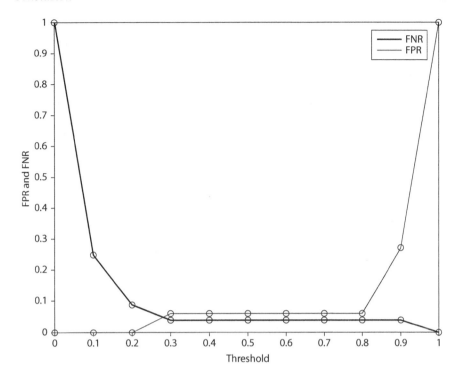

FIGURE 4.7 False positive and negative rates against different threshold values.

4.9, we can see that the TAN model could classify malware applications with a maximum accuracy of 0.98. The source code for extracting API, permission and system call based features are given in the Apendix. Here, Androguard tool [98] is used to extract API calls and permissions from the applications to construct csv files.

TABLE 4.9 Performance of the TAN-based model.

Dataset	TPR	FPR	Precision	Accuracy	F1Score
AMD+Drebin	0.98	0.04	0.96	0.97	0.96
Drebin	1	0.04	0.96	0.98	0.98

4.7 Conclusion

In this chapter, we discussed a mechanism for detecting Android malware applications by combining static and dynamic features related to the malicious activities by exploring their conditional dependencies. This hybrid detection mechanism can accurately capture the malicious behavior than existing static and dynamic analysis mechanisms. Further, this model is scalable to detect evolving malware and goodware applications. Evolving malware and goodware apps may use new features for performing their activities [154]. In such cases, the classifiers need to be retrained with the new set of features. Hence, by retraining, we can keep the TAN model up to date. In order to overcome the limitation of frequent retraining, one can employ active machine learning algorithms in the RRLR classifiers.

It is possible for an application to perform malicious activities by inheriting the permissions of other apps. In such a situation, the app can use the permissions declared in the manifest file of other apps. In this case, the AAPT tool cannot get these inherited permissions by analyzing the manifest file. Hence, a classifier trained with permission based features may wrongly classify the application. However, the API calls are still present in the source code of the application and after the execution of these API calls, malicious system calls will also be generated. Finally, the API call and the system call based classifiers can correctly classify the application. Hence, in most of the cases combining the outputs of permission, API call and system call based classifiers, we can detect the malicious behavior.

In this chapter, we discussed about the effectiveness of API calls, permissions and system calls for malware detection. API calls and permissions usually contain more information than system calls. However, the reliability of API calls and permissions can be questioned when a benign application invokes API calls and permissions that are typical of malware. Malware applications do not require any kind of user trigger for invoking sensitive API calls unlike goodware applications [101]. In goodware applications user triggers are propagated through intent mechanism for invoking sensitive APIs such as sendTextMessage() [101]. In those cases, the data reference is not send over the IPC (Inter Process Communication) channel [194]. However, the invocation of sensitive APIs without user triggers get reflected in the system call sequence generated by the application [211]. Hence, the system calls trace of an application contains relevant information about the malicious behavior than the standalone sensitive API calls invoked by the application.

Most of the existing system call-based dynamic analysis mechanisms rely on machine learning (ML) based approaches for detecting the malicious behavior. These ML based detection mechanisms take either system call frequencies [71] or co-occurrence matrix [209] or Markov chain state transition probability matrix [208] as the feature vector for the ML classifier. These approaches can result in high dimensionality of feature vectors [110]. Curse of dimensionality (high dimensionality of feature vectors) is a problem in which the higher number of features in a feature vector will result in high sparsity in the data [110]. This high sparsity in data can unnecessarily increase the storage space and processing time of the classifier. The limitations of existing system call-based dynamic analysis mechanisms are given in Table 4.10. In order to overcome these limitations, in the following chapters, we discuss low dimensional feature constructions using graph centrality measures and graph signals for effective malware detection. These graph based low dimensional features are then incorporated as part of machine learning algorithms that can automate the malware detection task.

TABLE 4.10   Limitations of existing system call-based mechanisms.

Approach	Limitations
Burguera et al. [89]	High dimensionality/Lack of system call dependencies
Xiao et al. [208]	High dimensionality
Xiao et al. [209]	High dimensionality
Zhang et al. [219]	High dimensionality
Bernadi et al. [83]	Requirement of multiple system call logs for feature construction
Yu et al. [216]	High dimensionality
Xiao et al. [210]	System call reordering attack
Canfora et al. [91]	High dimensionality
Bhandari et al. [85]	High dimensionality

5

Detection Using Graph Centrality Measures

In the Chapter 4, we showed that TAN model can be used to combine the static and dynamic features for accurate malware detection. In TAN-based detection model, it is possible for a malware developer to evade API call and permission-based classifiers by employing adversarial attacks[126].

Adversarial attacks on Android malware detection mechanisms are really threatening. They can make an attacker to gain unauthorized access to a device. In ML-based malware detection, feature values are extremely important since a slight change in the feature values can affect the output of the classifier. In addition to that, many adversarial attacks are transferable. That is the attacks targeting a specific ML-based classifier can also cause misclassifications in other ML-based classifiers. There are many simple feature manipulation techniques such as injecting bytes or appending bytes at the end of the application code. These simple feature manipulations can be easily detected by monitoring the file structure of the Android application. However, if an adversary adopts fine-grained modifications of the features, they can-not be easily detected.

Adversarial attacks may affect the performance of the TAN based detection model discussed in the Chapter 4. However, for an adversary it is not possible to change the system call sequence generated by an application without changing the underlying semantics in the source code of the application. Further, it is known that malware apps invoke sensitive APIs in an automated manner. This invocation of API calls without user triggers gets reflected in the corresponding system call sequence [211]. Hence, the system call sequence of an application contains relevant information about the malicious behavior than sensitive API calls. Hence system call based malware detection mechanisms are more robust to adversarial attacks. Many existing system call based detection mechanisms use standalone features related to the system call frequencies to detect the malicious behavior from the system call trace. The standalone features may not capture all the characteristics of system call trace associated with the malware application. Hence, the machine learning classifiers trained with these features can wrongly classify malware applications as goodware and vice versa. In order to overcome this limitation, a graph centrality based approach was proposed in [187]. Therefore, in this chapter, we discuss the computation of more informative features using graph centrality measures from system call traces for accurate malware detection.

DOI: 10.1201/9781003121510-5

5.1   Digraph from System Call Sequence

In Chapter 3, we explained the mechanism of tracing the system call sequence X of an application. A directed graph (digraph) is a convenient way of modelling the pairwise relationships between the elements in a complex sequence. The system call sequence X of an application can be represented as a directed graph $G = (S, \mathcal{E}, \mathcal{A})$, where $S = \{S_i : i = 1, 2, \ldots, n\}$ is the set of n system calls, A is the weighted adjacency matrix in which each $a_{ij} \in \mathcal{A}$ is the number of immediate occurrence of the system call S_j after the system call S_i in the system call sequence X and $\mathcal{E}$ is the set of m directed edges between the vertices in S. An edge e_{ij} exists from the vertex S_i to the vertex S_j if and only if $a_{ij} > 0$, that is if and only if in the execution of the application at least once the system call S_j occurs immediately after the system call S_i. The set of selected system calls S are given in Table 5.1.

Wang et al. [202], showed that memory management calls, error system calls and information maintenance system calls do not hold any significant impact on the malicious behavior. So, these system calls can be removed from the recorded system calls for effective identification of malicious behavior. Hence the system call sequence need to be refined by eliminating these system calls. The selected system calls are given in Table 4.5 (Chapter 4).

The refined system call sequence of the walkinwat trojan after executing in an emulator is shown in Figure 5.1. The system call logs can be collected using the strace utility. Here, we injected 1000 random events using monkeyrunner tool while collecting the system call sequence.

TABLE 5.1   List of relevant system calls.

Alternative Name	A	B	C	D	E	F	G	H
System Call	recvfrom	write	ioctl	read	sendto	dup	writev	pread
Alternative Name	I	J	K	L	M	N	O	P
System Call	close	socket	bind	connect	mkdir	access	chmod	open
Alternative Name	Q	R	S	T	U	V	W	X
System Call	rename	fchown32	unlink	pwrite	umask	lseek	fcntl	recvmsg
Alternative Name	Y	Z	A1	A2	A3	A4		
System Call	sendmsg	epoll	dup2	fchown	readv	chdir		

The system call digraph of the walkinwat trojan obtained from the system call sequence is shown in Figure 5.2 These system call graphs can capture the complex relationships among system calls in that system call sequence. After modelling the system call sequence as an ordered graph, we infer the relevant information related to the system calls in terms of centrality measures.

Z[AABACBCBDBDBBBDBDBBBDC]Z[DAAA]Z[C]Z[AC]ZZ[AABACBCBDBDBBBDBDBBBDC]Z[DAAA]Z[C]Z[AC]ZZ[AABACBCBDBDBBB
DBDBBBDC]Z[DAAA]Z[C]Z[AC]ZZ[AABACBCBDBDBBBDBDBBBDC]Z[DAAA]Z[C]Z[AC]ZZ[AABACBCBDBDBBBDBDBBBDC]Z[DAAA]
Z[C]Z[AC]ZZ[AABACBCBDBDBBBDBDBBBDC]Z[DAAA]Z[C]Z[AC]ZZ[AABACBCBDBDBBBDBDBBBDC]Z[DAAA]Z[C]Z[AC]ZZ[AABA
CBCBDBDBBBDBDBBBDC]Z[DAAA]Z[C]Z[AC]ZZ[AABACBCBDBDBBBDBDBBBDC]Z[DAAA]Z[C]Z[AC]ZZ[AABACBCBDBDBBBDBDBBB
DC]Z[DAAA]Z[C]Z[AC]ZZ[AABACBCBDBDBBBDBDBBBDC]Z[DAAA]Z[C]Z[AC]ZZ[AABACBCBDBDBBBDBDBBBDC]Z[DAAA]Z[C]Z[
AC]ZZ[AABACBCCBDBDBBBDBDBBBDCC]Z[DAAA]Z[C]Z[AC]ZZ[AABACBCBDBDBBBDBDBBBDC]Z[DAAA]Z[C]Z[AC]ZZ[AABACBCB
DBDBBBDBDBBBDC]Z[DAAA]Z[C]Z[AC]ZZ[AABACBCBDBDBBBDBDBBBDC]Z[DAAA]Z[C]Z[AC]ZZ[AABACBCBDBDBBBDBDBBBDC]Z
[DAAA]Z[C]Z[AC]ZZ[AABACBCBDBDBBBDBDBBBDC]Z[DAAA]Z[C]Z[AC]ZZ[AABACBCBDBDBBBDBDBBBDC]Z[DAAA]Z[C]Z[AC]Z
Z[AABACBCBDBDBBBDBDBBBDC]Z[DAAA]Z[C]Z[AC]ZZ[AABACBCCBDBDBBBDBDBBBDC]Z[DAAA]Z[C]Z[AC]ZZ[AABACBCBDBDBBB
BDBDBBBDC]Z[DAAA]Z[C]Z[AC]ZZ[AABACBCBDBDBBBDBDBBBDC]Z[DAAA]Z[C]Z[AC]ZZ[AABACBCBDBDBBBDBDBBBDC]Z[DAAA
]Z[C]Z[AC]ZZ[AABACBCBDBDBBBDBDBBBDC]Z[DAAA]Z[C]Z[AC]ZZ[AABACBCBDBDBBBDBDBBBDC]Z[DAAA]Z[C]Z[AC]ZZ[AAB
ACBCBDBDBBBDBDBBBDC]Z[DAAA]Z[C]Z[AC]ZZ[AABACBCBDBDBBBDBDBBBDC]Z[DAAA]Z[C]Z[AC]ZZ[AABACBCBDBDBBBDBDBB
BDC]Z[DAAA]Z[C]Z[AC]ZZ[AABACBCCBDBDBBBDBDBBBDCC]Z[DAAA]Z[C]Z[AC]ZZ[AABACBCCBDBDBBBDBDBBBDC]Z[DAAA]Z[
C]Z[AC]ZZ[AABACBCBDBDBBBDBDBBBDC]Z[DAAA]Z[C]Z[AC]ZZ[AABACBCBDBDBBBDBDBBBDC]Z[DAAA]Z[C]Z[AC]ZZ[AABACB
CCBDBDBBBDBDBBBDC]Z[DAAA]Z[C]Z[AC]ZZ[AABACBCBDBDBBBDBDBBBDCC]Z[DAAAACCCBDBDBBBDBDBBBDC]Z[AA]Z[A]Z[C]
Z[AC]ZZ[AABACBCBDBDBBBDBDBBBDCC]Z[DAAA]Z[C]Z[AACEA]ZZ[AABAACCBDBDBBBDBDBBBDCBCBDBDBBBDBDBBBDC]Z[DAAA
ACCBDBDBBBDBDBBBDC]Z[AAAEA]Z[A]Z[CCCCCC]Z[AC]Z[AACC]Z[BIBICBI]Z[CCIB]Z[DACCCBDBDBBBDBDBBBDC]Z[AA]ZZ[
GCC]Z[AACCCBDBDBBBDBDBBBDC]Z[AA]Z[CC]ZZZZZZZ[AACBIBICBICCBI]Z[AC]Z[AA]ZZZ[ACB]Z[DAAA]ZZ[AC]ZZ[AABACB
]Z[D]Z[AABAAC]Z[DAA]ZZ[ICCCCCC]Z[CCIGC]Z[A]Z[CC]ZZZ[CCI]ZZ[C]ZZZ[DCCCC]Z[C]Z[DGDDDDCCCACCCFI]Z[GDCCF
I]Z[CC]Z[AA]Z[AABACCCCC]Z[DAA]ZZ[AACCCCCCFBDIBDBBBDCCCCCFBDIBDBBBDCCC]Z[AC]Z[AA]ZZZ[CCCCFI]ZZ[GAAACC
CCCFBDIBDBDBBBDBDBBBDCCBCACCC]Z[DAA]ZZZZ[AACCCCCCCFBDIBDBBBDCCCCFBDIBDBDBBBDBDBBBDC]Z[AA]Z[AAACCCCCC
BDBDBBBDBDBBBDC]Z[AA]ZZ[AACCCBDBDBBBDBDBBBDCC]Z[AA]ZZ[AACCCBDBDBBBDBDBBBDC]Z[AA]Z[AACCCBDBDBBBDBDBBB
DC]Z[AA]Z[AACCBDBDBBBDBDBBBDCC]Z[AA]Z[AACCCBDBDBBBDBDBBBDCC]Z[AA]Z[A]Z[C]ZZ[AC]Z[AAAACBCCFBDIBDBDBBB
DBDBBBDCCBDBDBBBDBDBBBDCC]Z[DAAAEAAAA]Z[AAEEEAEAAACBCCCCFBDIBDBDBBBDBDBBBDCCBDBDBBBDBDBBBDC]Z[DAA]Z[C
]ZZ[AAACBCCBDBDBBBDBDBBBDCCBDBDBBBDBDBBBDC]Z[DAA]Z[CC]Z[BIBICCBI]Z[CCI]Z[AACCBDBDBBBDBDBBBDC]Z[AA]ZZ
[GCC]Z[AACCCBDBDBBBDBDBBBDCC]Z[AA]Z[CCCC]ZZZZZZZ[AACCBIBICCBICCCCBI]Z[AC]ZZ[AAACB]Z[DAAAAC]Z[AA]ZZZ[
A]ZZ[AC]Z[AAACB]Z[DAAA]ZZ[AC]Z[AAACB]Z[DAAA]ZZ[AC]ZZ[AABACB]Z[DAAA]ZZ[ICCCCCC]Z

FIGURE 5.1 System call sequence of Walkinwat Trojan.

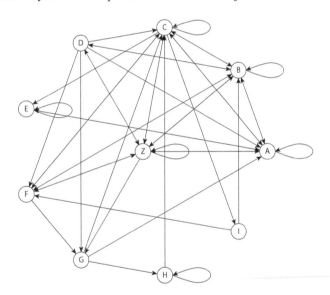

FIGURE 5.2 System call digraph of Walkinwat Trojan.

5.2 Centrality Measures from System Call Digraph

Centrality of a vertex is a measure of the influence of the vertex in the digraph G [157]. There exists different types of centrality measures such as, indegree centrality, eigen vector centrality, betweenness centrality, closeness centrality and so on. They are described below.

1. **Eigen Vector Centrality**: Let $\mathcal{N} = \{n_{ij} : i, j = 1, 2, 3, \ldots, n\}$ be a neighbourhood matrix of the graph G, where n_{ij} is computed as:

    $$n_{ij} = \begin{cases} 1, & \text{if } a_{ij} \neq 0; \\ 0, & \text{otherwise.} \end{cases}$$

    The eigen vector centrality of a vertex S_i is computed as:

    $$C_1(S_i) = \frac{1}{\lambda} \sum_{j=1}^{n} n_{ij} C_1(S_j), \tag{5.1}$$

    where λ is the largest eigen value of the adjacency matrix.

2. **Closeness Centrality**: It is the rate in which a vertex is closer to other vertices in the digraph. In a connected graph, the closeness centrality of a vertex S_i is calculated as:

    $$C_2(S_i) = \frac{n-1}{\sum_{j=1}^{n} \sigma_{ji}}, \tag{5.2}$$

    where σ_{ji} is the length of a shortest path from vertex S_j to vertex S_i and n is the number of vertices.

3. **Betweenness Centrality**: Betweenness centrality is the rate in which a vertex S_i lies in the shortest path between other vertices in the directed graph. The betweeness centrality $C_3(S_i)$ of a vertex S_i is computed as:

    $$C_3(S_i) = \sum_{j=1}^{n} \sum_{k=1}^{n} \frac{\sigma_{jik}}{\sigma_{jk}}, \tag{5.3}$$

    where σ_{jk} is the number of shortest paths from vertex S_j to vertex S_k and σ_{jik} is the number of shortest paths from vertex S_j to vertex S_k through vertex S_i.

4. **Indegree centrality**: It is the number of edges incident on a vertex of a directed graph G. The indegree centrality of a vertex S_i is calculated as:

    $$C_4(S_i) = \sum_{j=1}^{n} a_{ji}, \tag{5.4}$$

    where a_{ji} is the number of edges from the vertex S_j to vertex S_i.

The adjacency and the shortest distance matrices of the walkinwat trojan are shown in Figures 5.3 and 5.4, respectively. The indegree, eigen, betweenness and the closeness centrality values computed are given in Table 5.3.

	A	B	C	D	E	F	G	H	I	Z
A	189	41	114	0	4	0	0	0	0	68
B	41	238	41	303	0	14	0	0	0	7
C	2	109	126	0	1	5	0	0	10	153
D	49	232	62	0	0	7	1	0	0	1
E	5	0	0	0	2	0	0	0	0	0
F	0	7	0	0	0	3	0	0	0	0
G	1	0	3	0	0	0	0	2	0	0
H	1	0	3	0	0	0	0	2	0	0
I	0	12	8	0	0	0	1	0	0	4
Z	129	2	46	52	0	2	3	0	0	77

FIGURE 5.3 Adjacency matrix of walkinwat malware.

	A	B	C	D	E	F	G	H	I	Z
A	0	1	1	2	1	1	1	3	2	1
B	1	0	1	1	2	2	2	3	2	1
C	1	1	0	1	2	1	2	3	1	1
D	1	1	1	0	2	2	2	3	1	1
E	1	2	2	3	0	3	3	4	3	2
F	2	1	2	2	3	0	2	3	1	2
G	1	2	1	3	2	2	0	1	2	2
H	2	2	1	3	2	2	3	0	2	2
I	2	1	1	2	2	2	1	2	0	1
Z	1	1	1	1	2	2	1	2	1	0

FIGURE 5.4 Shortest distance matrix of walkinwat malware.

The code for representing a system call sequence as digraph and extracting the centrality measures such as eignvector centality, betweeness centrality and closeness centrality is given in the Appendix. Here, we used the nextworkx package [116] in python for representing the system call sequence as digraph and to extract the centrality measures.

TABLE 5.2 System call counts and their normalized values.

System Call (Alternative Name)	Count	Normalized Count
A	416	0.190650779
B	641	0.293767186
C	402	0.184234647
D	355	0.162694775
E	7	0.003208066
F	10	0.004582951
G	5	0.002291476
H	5	0.002291476
I	31	0.014207149
Z	310	0.142071494

TABLE 5.3 Centrality values of system calls in walkinwat digraph.

System call	Indegree Centrality	Eigen centrality	Betweenness centrality	Closeness centrality
A	0.19	0.4218	11.83	0.2700
B	0.29	0.4305	7.58	0.2700
C	0.18	0.4591	22.08	0.2945
D	0.16	0.1874	0.67	0.1705
E	0.003	0.1959	0	0
F	0.004	0.1021	0	0.1800
G	0.002	0.2120	8.25	0.1906
H	0.002	0.0471	0	0.1409
I	0.014	0.3538	5.50	0.2314
Z	0.014	0.4120	10.08	0.2492

5.3 Malware Detection Phase

The malicious behavior of applications can be detected from the centrality measures using ensemble learning method. Let $C_1 = (C_{1,1}, C_{1,2}, \ldots, C_{1,n})$ denotes the eigen vector centrality based feature vector of an application where $C_{1,i}$ for $i = 1, 2, \ldots, n$ denotes the eigen vector centrality of the system call S_i. Let $C_2 = (C_{2,1}, C_{2,2}, \ldots, C_{2,n})$ denotes the closeness centrality based feature vector of an application, where $C_{2,i}$ for $i = 1, 2, \ldots, n$ denotes the closeness centrality of the system call S_i. Let $C_3 = (C_{3,1}, C_{3,2}, \ldots, C_{3,n})$ denotes the betweeness centrality based feature vector of an application where $C_{3,i}$ for $i = 1, 2, \ldots, n$ denotes the betweenness centrality of the system call S_i.

Three ML classifiers can be trained with C_1, C_2, C_3 as feature vectors respectively. These three classifiers give estimate of the malicious scores from the centrality based feature vectors C_1, C_2 and C_3 of an unknown application. The malicious scores are further combined to detect whether the unknown app is a malware or not.

Let $O_i \in [0, 1]$ for $i = 1, 2, 3$ be the malicious scores produced by the three classifiers based on the centrality features C_1, C_2 and C_3 of the unknown application. The average malicious score O_{avg} of the application is calculated as,

$$O_{avg} = \sum_{i=1}^{3} w_i.O_i,$$

where $w_i \in [0, 1]$, for $i = 1, 2, 3$ are the weights of the classifiers and $\sum_{i=1}^{3} w_i = 1$. If the average malicious score O_{avg} is greater than a threshold T, the application can be treated as a malware.

5.4 Experiments and Analysis

In this section, we discuss about the performance of this detection mechanism. In Section 5.4.1, we discuss about the benign and malware datasets used. In Section 5.4.2, we discuss about the performance results and compare it with other approaches.

5.4.1 Dataset

In this section, we discuss about the dataset. We built a balanced dataset consisting of 2600 malicious and benign applications for demonstrating the performance of this mechanism. The malware applications were collected from Drebin [75] and AMD datasets [203] based on the code obfuscation techniques used and the type of the malware application. This dataset consists of 685 malware apps which use at least one obfuscation technique. The code obfuscation techniques employed by the obfuscated malware app in this dataset are the following:

- String renaming;

- Dynamic code loading;

- String encryption;

- Embedding native code.

The distribution of obfuscated malware applications are given in Table 5.4. The goodware apps are downloaded from google play store. All benign apps were submitted to virustotal [60] for verifying their legitimacy.

Android emulator can be used to collect the system call logs of all the malicious and benign applications in the dataset. In our experiments, the Android emulator was installed in an intel core i5 PC with 8GB memory. Here, we used strace tool to collect the system call trace of applications which were installed in the Android emulator

TABLE 5.4   Statistics of obfuscation techniques used in selected malware apps.

Obfuscation Technique	Number of Malware Apps
String renaming	435
Dynamic code loading	277
String encryption	507
Embedding native code	250

[185]. Monkey runner tool was used as an automated test case generation tool while tracing the system calls [153]. The monkey runner tool can get its maximum code coverage within a minute [82]. After one minute, we terminated the execution of the applications and saved the system call traces as log files. The collected system call logs were preprocessed by removing the arguments and eliminating the irrelevant system calls from them. From these preprocessed system call logs, we built a csv (comma separated value) file corresponding to eigen centrality, closeness centrality and betweenness centrality based features of 1300 benign and 1300 malicious applications.

5.4.2   Performance results

In this section, we discuss about the performance of the centrality based detection mechanism in the datasets described in Section 5.4.1. We used the features of 75% of malware and benign apps for training the classifiers and the features of 25% of malware and benign applications for testing the performance of the model. However, for measuring the performance in obfuscated malware apps, we used new ML models trained with the features of 75% malware (obfuscated and non-obfuscated) and goodware (samples used in previous model) and used the features of 25% of malware (obfuscated) and goodware (samples used in previous models) for testing. Here, we used ANN (Artificial Neural Network) classifier for testing the performance.

We treated the three centrality measures as three different feature vectors for three different ANN classifiers. So, we trained three ANN classifiers with closeness, eigen and betweeness centrality measures as features and combined their outputs for malware detection. The performance of individual ANN classifiers and their ensemble for various threshold values are given in Figure 5.5. From this figure we can conclude that the ensemble of classifiers gives better accuracy. Hence the ensemble of these three classifiers is used for the final prediction. Further, notice that the accuracy is highest for threshold values near 0.5. Hence, we fixed 0.5 as the threshold (malware-goodware margin) for all the classifiers.

For analyzing the performance of the model, we assigned various weights $w_i \in [0, 1]$ for $i = 1, 2, 3$, such that $w_1 + w_2 + w_3 = 1$ to the three classifiers for combining the malicious scores. The performance results are given in Table 5.5. From Table 5.5, we can see that ensemble averaging based mechanism can detect malware applications with good accuracy and F1 score while using artificial neural network as the ML classifier. Notice that the best performance with accuracy 98% occurs when the weight w_2 corresponding to the closeness centrality feature based classifier is given

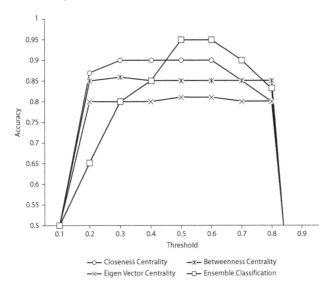

FIGURE 5.5 Performance results of single and ensemble of classifiers.

TABLE 5.5 Performance results with different weights to the classifiers.

Dataset	w_1	w_2	w_3	True Positive Rate	False Positive Rate	Positive Predictive Value	Accuracy	F1Score
All Malware Apps	$\frac{1}{3}$	$\frac{1}{3}$	$\frac{1}{3}$	0.95	0.03	0.97	0.96	0.96
All Malware Apps	0.30	0.40	0.30	0.97	0.02	0.98	0.97	0.97
All Malware Apps	0.25	0.50	0.25	0.98	0.02	0.98	0.98	0.98
All Malware Apps	0.25	0.45	0.35	0.98	0.02	0.98	0.98	0.98
Obfuscated Malware Apps	$\frac{1}{3}$	$\frac{1}{3}$	$\frac{1}{3}$	0.94	0.03	0.97	0.96	0.95
Obfuscated Malware Apps	0.30	0.40	0.30	0.96	0.02	0.98	0.97	0.96
Obfuscated Malware Apps	0.25	0.50	0.25	0.96	0.02	0.98	0.97	0.96
Obfuscated Malware Apps	0.25	0.45	0.35	0.96	0.02	0.98	0.97	0.96

higher value. Further, this mechanism can detect malware samples which use obfuscation techniques with an accuracy of 97%. Thus this mechanism can detect both the obfuscated and the non-obfuscated malware with almost the same accuracy. We also found that the ensemble averaging based mechanism is slightly more accurate than the majority voting based ensemble method (taking the majority of class output by the classifiers).

We compared the performance of this mechanism against various ML-based mechanisms which use other system call-based features such as system call frequency, transition probability matrix and system call graph signals [189]. Here, we extracted system call frequencies, transition probability matrix and graph signal-based features of 2600 apps in our dataset and used them as features of different ML classifiers. Here, we obtained the maximum accuracy with Random Forest (RF) classifier for all these feature vectors. The performance results are given in Table 5.6. From Tables 5.5 and 5.6, we can see that, the centrality feature based mechanisms outperform the other feature based mechanisms with better accuracy.

TABLE 5.6   Performance of RF classifier with various feature vectors.

Feature Vector	TPR	False Positive Rate	Positive Predictive Value	Accuracy	F1 Score
System Call Frequency Vector	0.91	0.14	0.86	0.88	0.88
System Call Transition Probability Matrix	0.89	0.11	0.89	0.89	0.89
Graph Signals	0.96	0.03	0.97	0.96	0.96

The program code for classification is given in the Appendix. We used R programming language for training and testing ANN classifiers. The training and testing files are given as inputs to the nnet() function and the prediction probabilities returned by the predict() function are averaged together. If the average probability value exceeds 0.5 then, the applications are treated as malware.

5.5   Conclusion

In this chapter, we discussed about the advantage of combining many types of features that can be inferred from the system call graph for malware detection. The graph centrality features provide better accuracy than standalone features used in many existing mechanisms. Further, this mechanism is capable of detecting malware apps which employ obfuscation techniques.

In our experiments, around 3% of goodware applications were wrongly classified as malware. This was because certain goodware apps try to request more privileges for their execution. Similarly, around 5% of malware apps could not be detected. This was because these malware apps did not fully exhibit their malicious behavior during the analysis time. To overcome these limitations one can try to infer more features from the system call digraph for enhancing the accuracy.

6

Graph Convolutional Network for Detection

Graph Convolutional Network (GCN) is a graph representation learning approach that represents the structure and features of the graph in a low dimensional Euclidean space. GCN has found to give promising results in many real-world applications such as learning social networks[206], traffic prediction[220], drug response prediction [156], etc. as well as in Android malware detection [128]. In this chapter, we discuss the application GCN in Android malware detection and illustrate with the detection of obfuscated Android malware from system call graphs. We organised the chapter as follows. Section 6.2 gives an introduction of GCN and its applications in many real-world problems, Section 6.3 explains GCN-based malware detection, Section 6.4 gives the results and discussions, Section 6.5 describes a case study and Section 6.6 gives the conclusions.

6.1 Introduction to GCN

Learning the structure of complex graphs is a challenging problem in many real-world applications[207]. To address this problem, graph representation learning approach is used in which the structure and features of a graph is represented in a low dimensional Euclidean space with embedding techniques. Although, graph embedding techniques give promising results in many applications, these mechanisms suffer from the limitations of shallow learning in which they fail to discover the complex structural complexities present in the graph. Although, deep learning mechanisms like CNN, RNN, etc. can solve the problems of shallow learning mechanisms, the non-Euclidean characteristics of graphs make convolutions and filtering ineffective [207]. Thus recently, a model called Graph Neural Network (GNN) has been proposed [88] that uses graph representation learning by utilizing the power of deep learning. To perform convolutions on the graph data, a variant of GNN called GCN has also emerged in the recent years [207][133]. GCN models are neural network models that can learn the graph structure and can aggregate the node information in a convolutional fashion. There are two types of GCN: spatial-based GCN and spectral based GCN[207]. In spatial GCN, the graph convolutions are defined by collecting the information from the nearby nodes as well as its own and in spectral GCN, convolutions are defined in the Fourier domain by computing the eigen decomposition of the graph Laplacian[207]. In the malware detection mechanism discussed in this chapter, we will be using spatial

DOI: 10.1201/9781003121510-6

GCN that aggregates the features of the system calls in the system call graph to obtain a compact feature representation of the Android applications. The following sections explain how GCN can be used effectively to detect Android malware.

6.2 GCN-Based Malware Detection

Malware developers can employ a variety of mechanisms to circumvent its detection by generating malware variants that mimic legitimate applications[196]. The common evasion strategies are repackaging, payload insertion, string encryption, etc. The ineffectiveness of antimalwares in detecting obfuscated variants has been widely discussed before[107]. In the year 2021, many malware variants have evaded the Google Play Protect mechanism[35]. The ineffectiveness of choosing static features for malware detection is widely discussed in the past[107]. Hence, inorder to cope up with the emerging malware threats, many malware detection mechanisms are using system call based detection to detect obfuscated Android malware[91]. The reason is that, all the requests made by the applications are passed through the system call interface before reaching the kernel of the device. Hence even if the applications are obfuscated, the system calls can be used to determine the malicious nature of the Android applications [109]. Many research works demonstrate the effectiveness of the structural dependencies between the system calls for detecting malware[122]. Hence in this chapter, we introduce an emerging deep learning approach called Graph Convolutional Networks[186], that is able to capture the interdependencies between the system calls in the system call graph. The main features of this approach are the following:

* A GCN based Android malware detection mechanism that uses the power of deep learning and system call dependency information to detect Android malware.

* A four dimensional feature representation with centrality values as features for classification that can better represent the behavior of Android malware.

* Ability to detect obfuscated malware variants.

The details of this mechanism are given in the following sections.

6.2.1 System call graph construction

Android kernel invocation calls can be classified into binder calls, system calls, and socket calls[50]. When an application requests the Android kernel to provide the required resources, the request is transferred to the system call interface. This interface serves as a layer between the user space and the kernel space. The system call is then executed in the kernel space and the control is passed to the user space. To capture the system calls, Android applications are made to run in an emulator. After that, one thousand pseudo random events such as touch event and key press event are injected to the applications using monkeyrunner tool.The system calls are then extracted using *strace* utility. After obtaining the system calls, the arguments are eliminated and then the system calls that are used for file management and network access are selected for constructing the graph. Table 6.1 shows the relevant system calls that were taken for developing the detection mechanism discussed in this chapter. The main objective of selecting these system calls is that malware use these system calls for accessing the sensitive resources in a device.

The system call sequence X of an application can be represented as a directed graph $G = (S, E)$, where $S = \{S_i : i = 1, 2, \ldots, n\}$ is the set of n relevant system calls, and E is the set of directed edges. An edge e_{ij} exists from the vertex S_i to the vertex S_j if the system call S_j occurs immediately after the system call S_i. It is important to be noted that the system call digraph has no multiple edges. Figures 6.1 and 6.2 shows the system call digraphs of Cerebrus malware and a benign Android application.

TABLE 6.1 Most relevant 26 system calls chosen.

Alternative Name	System Call	Description	Alternative Name	System Call	Description
A	recvfrom	system call that is used to recieve a message from a socket	B	write	system call that is used to write to a file descriptor
C	ioctl	system call that is used to manipulate the underlying device parameters	D	read	system call that is used for read operation
E	send to	system call that is used to send a message on a socket	F	dup	system call that is used to create a copy of the file descriptor
G	writev	system call that is used to write data to manipulate buffer	H	pread	system call that is used to write to or read from a file descriptor to a given offset
I	close	system call that is used to close a file descriptor	J	socket	system call that is used to create an endpoint for communication
K	bind	system call that is used to bind a name to a socket	L	connect	system call that is used to start a connection on a socket directory
M	mkdir	system call that is used to create a directory	N	access	system call that is used to check the users permission for accessing the file
O	chmod	system call that is used to change the permission of a file	P	open	system call that is used to open a file specified by the path name.
Q	fchown	used to change the ownership of a file.	R	rename	system call that is used to change the location or name of a file.
S	unlink	system call that is used to remove a file	T	pwrit	system call that is used to write or read from a file descriptor.
U	unmask	system call that is used to get file mode creation mask	V	fcntl64	system call that is used to change the file descriptor
W	recvmsg	system call that is used to recieve a message from a socket	X	sendmsg	system call that is used to send a message on a socket
Y	getdents64	system call that is used to obtain the directory entries	Z	epoll wait	system call that is used to wait for an I/O event

After constructing system call digraphs, we use the GCN to determine whether the digraph generated by the application corresponds to a malware or not. We use four graph centrality measures as features for detecting the malware. These centrality measures helps us to identify the most discriminating system calls that can be used for detecting the malware. In this chapter, we have taken the *Katz, Closeness, Betweenness* and *Page Rank* as different centrality measures. The details of these centrality measures are given below.

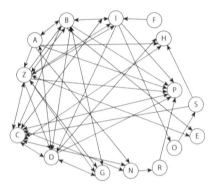

FIGURE 6.1 System call digraph of cerebrus malware disguised as covid tracker.

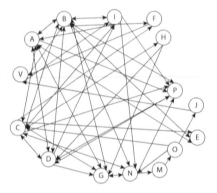

FIGURE 6.2 System call digraph of benign Android application.

1. Katz Centrality: This centrality value is used to indicate the influence of a vertex in the directed graph. Let $\mathcal{A}$ be the adjacency matrix of the directed graph $\mathcal{G} = (\mathcal{S}, \mathcal{E})$. The Katz centrality of the node $\mathcal{S}_i$ in the directed graph $\mathcal{G}$ is computed as,

$$K(\mathcal{S}_i) = \sum_{k=1}^{\infty} \sum_{j=1}^{n} \alpha^k \mathcal{A}_{ij}^k \qquad (6.1)$$

where $\mathcal{A}_{ij}^k$ is the $(i, j)^{th}$ entry of $\mathcal{A}^k$ which is equal to the total number of paths of length k between the vertices $\mathcal{S}_i$ and $\mathcal{S}_j$ and α represents the attenuation factor used to penalize the distance between them.

2. Closeness Centrality: It is the rate in which a vertex is closer to other vertices in the digraph. In a connected graph, the closeness centrality of a

vertex S_i is calculated as,

$$C_2(S_i) = \frac{n-1}{\sum\limits_{j=1}^{n} \sigma_{ji}}, \tag{6.2}$$

where σ_{ji} is the length of a shortest path from vertex S_j to vertex S_i and n is the number of vertices.

3. Betweenness Centrality: Betweenness centrality is the rate in which a vertex S_i lies in the shortest path between other vertices in the directed graph. The betweeness centrality $C_3(S_i)$ of a vertex S_i is computed as,

$$C_3(S_i) = \sum_{j=1}^{n} \sum_{k=1}^{n} \frac{\sigma_{jik}}{\sigma_{jk}}, \tag{6.3}$$

where σ_{jk} is the number of shortest paths from vertex S_j to vertex S_k and σ_{jik} is the number of shortest paths from vertex S_j to vertex S_k through vertex S_i.

4. Page Rank: Page rank centrality value finds the importance of a vertex in the directed graph whose influence extends beyond direct connections. The page rank of the vertices S_i are computed iteratively. Initially the page rank of all the vertices will be 1. At each iteration, the page rank of the vertex S_i will be updated as,

$$P(S_i) = (1 - \beta) + \beta\left(\frac{P(S_r)}{d(S_r)} + \cdots + \frac{P(S_t)}{d(S_t)}\right), \tag{6.4}$$

where β is the damping factor which is taken as 0.85, $P(S_r), \ldots, P(S_t)$ are the page ranks of the vertices $S_r, \ldots, S_t \in S$ that are pointing to the vertex S_i and $d(S_r), \ldots, d(S_t)$ are their out degrees.

If any of the system calls are absent in an application, we represent it as an isolated node in the digraph with all centrality values equal to 0.

6.2.2   GCN for low dimensional feature representation

The b centrality values of the n vertices in the system call graph gives an $n \times b$ feature matrix representation for the Android application. GCN can be used to convert this $n \times b$ feature matrix to a low dimensional feature matrix of size $l \times m$, where $l \leq n$ and $m \leq b$ for malware detection. Since the system call digraph has 26 vertices, with 4 centrality values for each of the vertices, the value of n is 26 and b is 4 in our case. GCN converts this 26×4 feature matrix to a 1×4 feature matrix. Hence with GCN, we will get a 4 dimensional feature representation for Android applications. Figure 6.3 shows the various steps in this method. The low dimensional feature representation is obtained using GCN as follows.

At the tth layer of the GCN, a vertex $S_i \in S$ has a hidden vector $h_{S_i}^t$, where the centrality values $h_{S_i}^0 = (K(S_i), C_2(S_i), C_3(S_i), P(S_i))$ denotes the initial values of the

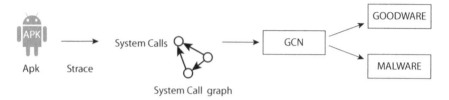

FIGURE 6.3 Proposed method.

hidden vectors. These hidden vectors are the *Katz, Closeness, Betweenness* and *Page Rank* centrality values of the vertex S_i. At the tth layer, the vertex S_i collects the information from the neighbouring vertices. It then computes the hidden vector $h^t_{S_i}$ of the node S_i as,

$$\tilde{h}^t_{S_i} = \sum_{u \in N(S_i) \cup \{S_i\}} h^{t-1}_u \tag{6.5}$$

$$h^t_{S_i} = \sigma(\tilde{h}^t_{S_i} W^t), \tag{6.6}$$

where $N(S_i)$ denotes the set of nodes that are connected to node S_i (neighbouring nodes). W^t denotes the weight matrix of size $d_{t-1} \times d_t$, where d_t denotes the size of the hidden vectors at the tth layer, and σ denotes the sigmoid activation function. In this manner, the hidden vectors are computed for all nodes in every layer. Finally, the hidden vector at the last layer (k^{th} layer) is the node embedding vector of the node S_i which is denoted as,

$$z_{S_i} = h_{S_i k} \tag{6.7}$$

After computing the hidden vector at the last layer, the node embedding vectors of all the nodes are added together to form the pooling layer which gives the graph representation z_G. That is,

$$z_G = \sum_{S_i \in S} z_{S_i}, \tag{6.8}$$

where z_{S_i} denotes the node embedding vector of node S_i.

6.2.3   Training of GCN

In the training phase, the input of the GCN will be the adjacency matrix and the feature matrix of the system call graphs of benign and malware applications. Let $Tr = \{(\mathcal{A}_i, N_i, Y_i) : i = 1, \ldots, p\}$ be the training set of p graphs (applications), where $\mathcal{A}_i$ denotes the adjacency matrix, N_i denotes the feature matrix which is the matrix of centrality values of nodes and lb_i denotes the label of the ith graph $G_i = (S_i, \mathcal{E}_i)$. The optimal weights W are computed as follows. First the weights are initialized randomly. Then the GCN is made to predict the label of the training sample $(\mathcal{A}_i, N_i)$, which is then compared with the actual label lb_i. Then the error or loss function is

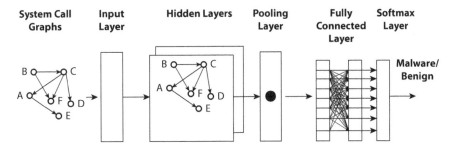

FIGURE 6.4   Malware detection with GCN.

calculated as the difference between the predicted label and the actual label Y_i. The weight matrix can be represented as a matrix W of size $d_{t-1} \times d_t$ where d_{t-1} is the dimension of the hidden vector at layer $t - 1$ and d_t is the dimension of the hidden vector at layer t. We can the use Adam optimization [132] to update the weights.

6.2.4   System call graph classification using GCN

To classify an Android application as malware or goodware, a fully connected layer with a softmax function is added after the pooling layer of the GCN. The function of this layer is to take the graph representation z_G from the pooling layer and to provide the probability of application to be a malware. Figure 6.4 shows the architecture of GCN.

6.3   Experiments and Analysis

To conduct the experiments, we collected malware samples from Drebin[75], AMD [203] and Malgenome [222]. The details of the malware families collected are given Table 6.2. The benign applications were downloaded from Google Play Store and uploaded to VirusTotal for verification. For building the model, a total of 1560 samples were taken. Among them, there were 560 malware samples and 1000 goodware samples [165]. A system with 64 bit Windows 10 operating system and Intel(R) Core(TM) i7-6700K CPU @ 4.00GHz with 32 GB RAM is used to carry out the experiments.

For collecting the system calls, the Android applications were made to run in Android Genymotion emulator. Monkeyrunner tool was used to inject pseudorandom events to an application and the system calls were collected with the *strace* utility. The system call sequences were then used to generate the system call graphs. After constructing the graphs, the centrality measures were computed and provided as node attributes to the graph. The adjacency matrices and the centrality values were then used for training the GCN model. The experimental results show that the four centrality values can distinguish goodware from malware.

TABLE 6.2   Android malware families.

Android Malware Family	Type of Obfuscation	Malware Type
Andup	String encryption and renaming	Adware
Boxer	String encryption and renaming	Trojan-SMS
DroidKungFu	Native payload, string encryption ,renaming	Backdoor
FakeAngry	String encryption and renaming	Backdoor
FakePlayer	Renaming	Trojan-SMS
Mmarketpay	-	Trojan
Lotoor	Dynamic loading and renaming	HackerTool
Kuguo	Renaming	Adware
Kyview	Renaming, string encryption	Adware
Lnk	-	Trojan
Mecor	-	Trojan-spy
Minimob	Renaming	Adware
Zitmo	Renaming	Trojan-Banker
Nandrobox	-	Trojan
RuMMS	Renaming, string encryption, dynamic loading	Trojan-SMS
Winge	Renaming	Trojan-Clicker
Penetho	-	HackerTool
Mseg	Renaming	Trojan
BrainTest	Dynamic loading	Backdoor
Stealer	Renaming	Trojan-SMS
FakeDoc	Renaming	Trojan
Tesbo	Renaming, string encryption, dynamic loading	Trojan-SMS
Mtk	Renaming, string encryption,dynamic loading	Trojan
Utchi	Renaming	Adware
Cerebrus	Java reflection	Banker

Table 6.3 shows the Katz centrality values of 6 system calls. This shows that even with Katz centrality values of a few system calls one may be able to distinguish malware from benign applications.

6.3.1   Implementation details

From 1560 application samples, we took 1248 samples for training, 156 samples for testing and 156 samples for validation. In the GCN model, we chose three hidden layers. Among the hidden layers, two were graph convolutional layers and the third was a single pooling layer. The accuracy values associated with the size of the layers are given in the Table 6.4. We obtained the maximum accuracy with hidden layers of size 26,11 and 4 respectively. We used metrics such as accuracy, precision, recall and F1 measure to evaluate the performance of the model. True Positives (TP) indicates the number of malware applications that are correctly classified as malware, True Negatives (TN) indicates the number of goodware applications that are correctly classified as goodware, False Positives (FP) indicates the number of benign applications that are incorrectly classified as malware, and False Negatives (FN) indicates the number of malware applications that are incorrectly classified as benign. The accuracy and

TABLE 6.3 Katz centrality values of selected system calls.

System Call	Malware	Benign
fcntl64	0.3361	0.1824
open	0.2134	0.1674
access	0.2321	0.2187
write	0.3657	0.2402
pread	0.2844	0.2621
sendto	0.3192	0.2141

F1-measure of the model are computed as given below.

$$Accuracy = \frac{TP + TN}{TP + FN + TN + FP}$$

$$Precision = \frac{TP}{TP + FP}$$

$$Recall = \frac{TP}{TP + FN}$$

$$F1 - measure = \frac{2 \times Precision \times Recall}{Precision + Recall}$$

The GCN classifier is trained with 100 epochs and a learning rate of 0.01. We used Adam optimization with an early stopping value of 10.

TABLE 6.4 Accuracy with size of hidden layers of GCN.

Size of Hidden Layers	Accuracy
(26,11,4)	0.913
(32,32,16)	0.873
(32,16,4)	0.853

Table 6.5 shows the performance of this mechanism with AMD, Drebin datasets and some latest malware families that use obfuscation(AMD 2020)[203]. We can see that the mechanism has an accuracy of 91.30%.

TABLE 6.5 Performance results.

TPR	FPR	FNR	TNR	Precision	Recall	F1 Score	Accuracy
0.912	0.086	0.088	0.914	0.913	0.912	0.912	0.913

To evaluate the effectiveness of the GCN based malware detection mechanism, we compared it with the existing system call frequency-based mechanism proposed by Fei et al[197] with Malgenome dataset, and system call frequency-based detection mechanisms using SVM (Support Vector Machines) as well as decision tree with Drebin datasets. The Malgenome dataset is now a part of the Drebin dataset [222]. Table 6.6 shows the comparison results. We found that with GCN, we can classify malware with 93.3% accuracy. For implementing SVM and decision tree, we used WEKA 3.8 with 10 fold cross validation.

TABLE 6.6   Comparison of the GCN method with other methods.

Method	TPR	FPR	FNR	TNR	Precision	Recall	F1 Score	Accuracy	Datasets
GCN Approach	0.938	0.071	0.062	0.929	0.929	0.938	0.933	0.933	Drebin
Fei et al.[197]	-	-	-	-	-	-	-	0.901	Malgenome
SVM	0.873	0.137	0.863	0.127	0.868	0.864	0.873	0.868	Drebin
Decision Tree	0.890	0.084	0.110	0.916	0.913	0.8907	0.901	0.903	Drebin

6.4   Detection of Emerging Malware

Android malware continue to emerge day by day and hence it is challenging to detect the malware in an effective and scalable manner[212]. Hence in this section, we explore whether GCN can detect the evolving Android malware.

TABLE 6.7   Malware families.

Malware	Number of Samples	Type	Year
MystryBot	5	Ransomware, Keylogger	2018
SMS malware	15	Spywares	2020
Banking malware	20	Trojan	2019
Comebot	5	Banking Spyware	2019
Descarga	10	Banking Trojan	2016
Xbot	1	Ransomware	2016
Covidlures	3	Spyware	2020

To test whether the GCN mechanism can detect the latest Android malware applications, we prepared a test set with the malware samples given in Table 6.7. The malware samples were collected from the public repositories[9],[8]. Figure 6.5 shows the system call graph of Xbot malware[12]. Xbot malware is the successor of a Trojan named Aulrin that appeared in 2014. After its installation in an Android device, this malware starts communicating with a command-and-control server (C2 server). The malware requests permission called RECEIVE_BOOT_COMPLETED that allows the malware to be persistent on the compromised device. It then steals the credit card information of the user. After that, it encrypts the files of the user. To evade de-tection, this malware uses a variety of obfuscation. From the system call graph, we can see that there exists connections between *writv()*, *socket()*, *sendmsg()*, *recvfrom()*, *and connect()* system calls, that are used to establish a connection to a C2 server to en-crypt the files. The performance results are given in Table 6.8. From this performance results, we can conclude GCN with centrality measures can detect many emerging malware applications.

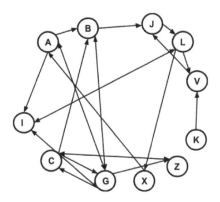

FIGURE 6.5   System call graph of Xbot malware.

TABLE 6.8   Detection rate of emerging malware.

TPR	FPR	FNR	TNR	Precision	Recall	F1 Score	Accuracy
0.890	0.084	0.110	0.916	0.913	0.8907	0.901	0.903

The code of GCN is given in the Appendix. Packages such as Keras and Tensor-flow are used to code GCN. We used the implementation of[87] to build our GCN model.

6.5   Conclusion

In this chapter, we presented a malware detection mechanism with system call graphs using GCN. This mechanism was able to detect Android malware with an accuracy of 91.3% in our dataset. We also explored whether GCN was able to detect the evolving Android malware. We obtained an accuracy of 90% on a dataset of latest Android malware applications. The ideas presented in this chapter will help cyber security researchers and developers to unveil the potential of GCN in developing automated cyber security solutions.

7

Graph Signal Processing-Based Detection

In the previous chapters, we discussed about the effectiveness of system call graph-based malware detection mechanisms. These mechanisms considered the vertex level features such as centralities for malware detection. In the current settings, it is very difficult to incorporate high dimensional edge level features such as adjacency matrix in a machine learning classifier. In order to overcome this limitation, a graph signal processing based approach was proposed in [189]. In this chapter, we first discuss how to generate various graph signals from system call sequence of an application for constructing low dimensional feature vectors. Later we discuss about employing various machine learning classifiers for effective malware detection using these feature vectors.

7.1 Graph Signal Processing and Its Applications

The information in a dataset with n related elements can be represented as a graph $\mathcal{G}=(\mathcal{S}, \mathcal{E}, \mathcal{A})$, where, $\mathcal{S} = \{S_1, \ldots, S_n\}$ is the vertex set consisting of elements in the dataset, $\mathcal{E}$ is the set of edges and $A = (a_{ij})_{n \times n}$ is a weighted adjacency matrix in which each a_{ij} denotes the degree of relationship between the vertices S_i and S_j.

Definition 7.1. *A graph signal $\mathcal{V}$ is a function from the set of vertices $\mathcal{S}$ into the set of real numbers $\mathbb{R}$ or complex numbers $\mathbb{C}$. That is,*

$$\mathcal{V} : \mathcal{S} \to \mathbb{R} \ or \ \mathbb{C}.$$

In this chapter, we will treat the graph signal $\mathcal{V}$ on S as a column vector. That is,

$$\mathcal{V}(S) = [\mathcal{V}(S_1) \ldots \mathcal{V}(S_n)]^T.$$

For convenience, we will denote $\mathcal{V}(S)$ as $\mathcal{V}$, wherever the vertex set $\mathcal{S}$ is obvious.

There exists a plethora of engineering and scientific applications where signals naturally exist on graph representations such as gene-expression patterns defined on gene networks, number of infections on the network of spread of an epidemic, rumors, or memes over a population network, congestion level at the nodes of a telecommunication network and so on. In all these cases, complex systems are formed by multiple nodes, where the global network behavior is a result of the local interactions between connected nodes. In many cases, our object of interest will be the information defined

DOI: 10.1201/9781003121510-7

on top of the graph, that is the information associated with the nodes of the graph. Graph signal processing (GSP) can be applied to extract this information by modeling the structure of the data using a graph and then viewing the available information as a signal defined on it. In GSP, a signal is first defined on each node of the graph. These signals are then processed using various connectivity relationships in the graph. The resulting signal values can provide better insights into the data represented by the graph. The graph shift is such an operation.

Definition 7.2. *The graph shift of a signal $\mathcal{V}$ on a vertex set S is obtained by replacing the signal $\mathcal{V}(S_i)$ at each vertex by a weighted linear combination of signals at the neighbors of S_i using the weighted adjacency matrix A. That is, the graph shift $\widetilde{\mathcal{V}}$ of $\mathcal{V}$ is calculated as,*

$$\widetilde{\mathcal{V}} = \mathcal{A}\mathcal{V}.$$

That is,

$$\widetilde{\mathcal{V}(S)} = \mathcal{A}[\mathcal{V}(S_1) \dots \mathcal{V}(S_n)]^T.$$

GSP has found tremendous applications in various fields such as brain disease prediction [161], social network analysis [121], weather prediction [175] and so on. In brain disease prediction [161], human activity signals are mapped into the regions (nodes) of the human brain network. These signals are then processed using the connectivity information between the brain regions (nodes). The processed signals are then used to identify the anomalies or diseases such as Alzheimer's in the human brain. In social networks analysis [121], the information about rumors (signals) are mapped onto the people (nodes) in the social network. These signals are then processed using the relationship information among the people. The processed signals are then used for finding the source of the rumor. In weather prediction [175], the temperature information obtained from various sensors (nodes) are taken as the signals. These signals are then processed using the distance information between the sensors. The processed signals are then used to analyze the relative temperature distribution across a geographical area. In the next section, we will discuss about representing the system call sequence of an Android application as graph signals.

7.2   Graph Signals from System Call Sequence

In this section, we discuss about representing system call sequence as a digraph and computing graph signals for each node. It consists of two steps. In the first step, we extract the system call sequence generated by an application and construct a digraph as described in Chapter 5. In the second step, we construct appropriate signals on the vertices of this system call graph.

The system call sequence of the Walkinwat trojan is shown in Figure 7.1 and the system call digraph is shown in Figure 7.2. The set of selected system calls are given in Table 7.1. The system call frequencies are given in Table 7.2 and adjacency matrix is given in Figure 7.3.

In the second step, we first define a signal on the vertices of this system call digraph as the normalized frequency of occurrence that system call. Then, we refine these signals by applying a graph shift operation (adjacency matrix) on these signals to obtain the graph signals with information on system call dependencies.

```
Z[AABACBCBDBDBBBBDBDBBBBDC]Z[DAAA]Z[C]Z[AC]ZZ[AABACBCBDBDBBBBDBDBBBBDC]Z[DAAA]Z[C]Z[AC]ZZ[AABACBCBDBDBBBB
DBDBBBDC]Z[DAAA]Z[C]Z[AC]ZZ[AABACBCBDBDBBBBDBDBBBDC]Z[DAAA]Z[C]Z[AC]ZZ[AABACBCBDBDBBBBDBDBBBDC]Z[DAAA]
Z[C]Z[AC]ZZ[AABACBCBDBDBBBBDBDBBBDC]Z[DAAA]Z[C]Z[AC]ZZ[AABACBCBDBDBBBBDBDBBBDC]Z[DAAA]Z[C]Z[AC]ZZ[AABA
CBCBDBDBBBBDBDBBBDC]Z[DAAA]Z[C]Z[AC]ZZ[AABACBCBDBDBBBBDBDBBBDC]Z[DAAA]Z[C]Z[AC]ZZ[AABACBCBDBDBBBBDBDBBB
DC]Z[DAAA]Z[C]Z[AC]ZZ[AABACBCBDBDBBBBDBDBBBDC]Z[DAAA]Z[C]Z[AC]ZZ[AABACBCBDBDBBBBDBDBBBDC]Z[DAAA]Z[C]Z[
AC]ZZ[AABACBCCBDBDBBBBDBDBBBDCC]Z[DAAA]Z[C]Z[AC]ZZ[AABACBCBDBDBBBBDBDBBBDC]Z[DAAA]Z[C]Z[AC]ZZ[AABACBCB
DBDBBBBDBDBBBDC]Z[DAAA]Z[C]Z[AC]ZZ[AABACBCBDBDBBBBDBDBBBDC]Z[DAAA]Z[C]Z[AC]ZZ[AABACBCBDBDBBBBDBDBBBDC]Z
[DAAA]Z[C]Z[AC]ZZ[AABACBCBDBDBBBBDBDBBBDC]Z[DAAA]Z[C]Z[AC]ZZ[AABACBCBDBDBBBBDBDBBBDC]Z[DAAA]Z[C]Z[AC]Z
Z[AABACBCBDBDBBBBDBDBBBDC]Z[DAAA]Z[C]Z[AC]ZZ[AABACBCBDBDBBBBDBDBBBDC]Z[DAAA]Z[C]Z[AC]ZZ[AABACBCBDBDBBBB
BDBDBBBDC]Z[DAAA]Z[C]Z[AC]ZZ[AABACBCBDBDBBBBDBDBBBDC]Z[DAAA]Z[C]Z[AC]ZZ[AABACBCBDBDBBBBDBDBBBDC]Z[DAAA
]Z[C]Z[AC]ZZ[AABACBCBDBDBBBBDBDBBBDC]Z[DAAA]Z[C]Z[AC]ZZ[AABACBCBDBDBBBBDBDBBBDC]Z[DAAA]Z[C]Z[AC]ZZ[AAB
ACBCBDBDBBBBDBDBBBDC]Z[DAAA]Z[C]Z[AC]ZZ[AABACBCBDBDBBBBDBDBBBDC]Z[DAAA]Z[C]Z[AC]ZZ[AABACBCBDBDBBBBDBDBB
BDC]Z[DAAA]Z[C]Z[AC]ZZ[AABACBCCBDBDBBBBDBDBBBDCC]Z[DAAA]Z[C]Z[AC]ZZ[AABACBCCBDBDBBBBDBDBBBDC]Z[DAAA]Z[
C]Z[AC]ZZ[AABACBCBDBDBBBBDBDBBBDC]Z[DAAA]Z[C]Z[AC]ZZ[AABACBCBDBDBBBBDBDBBBDC]Z[DAAA]Z[C]Z[AC]ZZ[AABACB
CCBDBDBBBBDBDBBBDC]Z[DAAA]Z[C]Z[AC]ZZ[AABACBCBDBDBBBBDBDBBBDCC]Z[DAAAACCCBDBDBBBBDBDBBBDC]Z[AA]Z[A]Z[C]
Z[AC]ZZ[AABACBCBDBDBBBBDBDBBBDCC]Z[DAAA]Z[C]Z[AECEA]ZZ[AABAACCBDBDBBBBDBDBBBDCBCBDBDBBBBDBDBBBDC]Z[DAAA
ACCBDBDBBBBDBDBBBDC]Z[AAAEA]Z[A]Z[CCCCCC]Z[AC]Z[AACC]Z[BIBICBI]Z[CCIB]Z[DACCCBDBDBBBBDBDBBBDC]Z[AA]ZZ[
GCC]Z[AACCCBDBDBBBBDBDBBBDC]Z[AA]Z[CC]ZZZZZZZ[AACBIBICBICCBI]Z[AC]Z[AA]ZZZ[ACB]Z[DAAA]ZZ[AC]ZZ[AABACB
]Z[D]Z[AABAAC]Z[DAA]ZZ[ICCCCCC]Z[CCIGC]Z[A]Z[CC]ZZZ[CCI]ZZ[C]ZZZ[DCCCC]Z[C]Z[DGDDDDCCCACCCFI]Z[GDCCF
I]Z[CC]Z[AA]Z[AABACCCCC]Z[DAA]ZZ[AACCCCCCFBDIBDBBBDCCCCCFBDIBDBBBDCCC]Z[AC]Z[AA]ZZZ[CCCCFI]ZZ[GAAACC
CCCFBDIBDBDBBBBDBDBBBDCCBCACCC]Z[DAA]ZZZZ[AACCCCCCCFBDIBDBBBDCCCCFBDIBDBDBBBBDBDBBBDC]Z[AA]Z[AAACCCCCC
BDBDBBBDBDBBBDC]Z[AA]ZZ[AACCCBDBDBBBBDBDBBBDCC]Z[AA]ZZ[AACCCBDBDBBBBDBDBBBDC]Z[AA]Z[AACCCBDBDBBBBDBDBBB
DC]Z[AA]Z[AACCBDBDBBBBDBDBBBDCC]Z[AA]Z[AACCCBDBDBBBBDBDBBBDCC]Z[AA]Z[A]Z[C]ZZ[AC]Z[AAAACBCCFBDIBDBDBBBB
DBDBBBDCCBDBDBBBBDBDBBBDCC]Z[DAAAEAAAA]Z[AAEEEAEAAACBCCCCFBDIBDBDBBBBDBDBBBDCCBDBDBBBBDBDBBBDC]Z[DAA]Z[C
]ZZ[AAACBCCBDBDBBBBDBDBBBDCCBDBDBBBBDBDBBBDC]Z[DAA]Z[CC]Z[BIBICCBI]Z[CCI]Z[AACCBDBDBBBBDBDBBBDC]Z[AA]ZZ
[GCC]Z[AACCCBDBDBBBBDBDBBBDCC]Z[AA]Z[CCCC]ZZZZZZZ[AACCBIBICCBICCCCBI]Z[AC]ZZ[AAACB]Z[DAAAAC]Z[AA]ZZZ[
A]ZZ[AC]Z[AAACB]Z[DAAA]ZZ[AC]Z[AAACB]Z[DAAA]ZZ[AC]ZZ[AABACB]Z[DAAA]Z[ICCCCCC]Z
```

FIGURE 7.1 System call sequence of Walkinwat Trojan.

TABLE 7.1 List of relevant system calls.

Alternative Name	A	B	C	D	E	F	G	H
System Call	recvfrom	write	ioctl	read	sendto	dup	writev	pread
Alternative Name	I	J	K	L	M	N	O	P
System Call	close	socket	bind	connect	mkdir	access	chmod	open
Alternative Name	Q	R	S	T	U	V	W	X
System Call	rename	fchown32	unlink	pwrite	umask	lseek	fcntl	recvmsg
Alternative Name	Y	Z	A1	A2	A3	A4		
System Call	sendmsg	epoll	dup2	fchown	readv	chdir		

Graph signal processing helps to analyze the structure of a system call connectivity digraph. We first consider the frequency value $F(S_i)$ of the system call S_i for $i = 1, \ldots, n$ in the system call digraph G for constructing the graph signals. The frequency $F(S_i)$ of the occurrence of each system call S_i is divided by the total number of system calls $\sum_{j=1}^{n} F(S_j)$. This normalized frequency values (probability) of system calls are taken as the initial signal values. In other words the probability of occurrence of each system call is taken as the initial signal values. The information about the rate of occurrence of each system call S_i in a system call sequence X can be inferred from

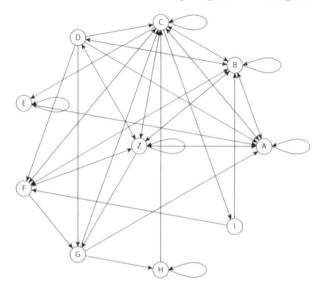

FIGURE 7.2   System call digraph of Walkinwat malware application.

TABLE 7.2   System call frequency values.

System Call Name	System Call (Alternative Name)	Frequency
recvfrom	A	416
write	B	641
ioctl	C	402
read	D	355
sendto	E	7
dup	F	10
writev	G	5
pread	H	5
close	I	31
epoll	Z	310

its system call sequence. The graph signal V_0 on S_i is defined as,

$$V_0(S_i) = \frac{F(S_i)}{\sum\limits_{j=1}^{n} F(S_j)} \qquad (7.1)$$

The adjacency value a_{ij} from a system call S_i to another system call S_j in the adjacency matrix A of the system call digraph is calculated as the number of occurrence of the system call S_j after the system call S_i, $\forall 1 \leq j \leq n$ in the system call sequence.

	A	B	C	D	E	F	G	H	I	Z
A	189	41	114	0	4	0	0	0	0	68
B	41	238	41	303	0	14	0	0	0	7
C	2	109	126	0	1	5	0	0	10	153
D	49	232	62	0	0	7	1	0	0	1
E	5	0	0	0	2	0	0	0	0	0
F	0	7	0	0	0	3	0	0	0	0
G	1	0	3	0	0	0	0	2	0	0
H	1	0	3	0	0	0	0	2	0	0
I	0	12	8	0	0	0	1	0	0	4
Z	129	2	46	52	0	2	3	0	0	77

FIGURE 7.3   Adjacency matrix of Walkinwat malware graph.

The normalized adjacency value $\widehat{a_{ij}}$ from a system call S_i to another system call S_j is calculated as,

$$\widehat{a_{ij}} = \frac{|a_{ij}|}{\sum\limits_{k=1}^{n} |a_{ik}|}, \forall 1 \le i, j \le n \tag{7.2}$$

The graph signal V_0 can be transformed to other signal V by various graph signal processing operations. We transform the graph signals to V using the normalized adjacency matrix $\widehat{A} = (\widehat{a_{ij}})$ as,

$$V = \widehat{A}V_0.$$

The graph signals and transformed signals of Walkinwat malware app is given in Table 7.3.

7.3   Machine Learning Classification for Malware Detection

The graph signals or processed signals obtained from the system call digraph of Android applications can be used as feature vectors of a supervised machine learning classifier for malware detection. Here, any machine learning algorithm such as naive Bayes, support vector machine (SVM), decision trees, etc. can be used for classifying the graph signals to detect the malicious behavior.

TABLE 7.3   Graph signals of Walkinwat Trojan.

System Call Name	System Call (Alternative Name)	$\mathcal{V}_0$	$\mathcal{V}_1$
recvfrom	A	0.190650779	0.1969
write	B	0.293767186	0.2181
ioctl	C	0.184234647	0.1985
read	D	0.162694775	0.2067
sendto	E	0.003208066	0.1689
dup	F	0.004582951	0.1886
writev	G	0.002291476	0.1616
pread	H	0.002291476	0.1486
dup	I	0.014207149	0.1863
epoll	Z	0.142071494	0.1909

7.3.1   Construction of low-dimensional feature vectors

Let $\mathcal{V}(S)$ denotes the signal vector of an Android application. Then we can have the n-dimensional feature vector as,

$$[\mathcal{V}(S_1)\dots\mathcal{V}(S_n)]^T.$$

One can reduce the feature dimension by using the signal values of only a few selected system calls as the feature vector. The system calls can be selected by analysing their signal values in malware and goodware applications. The signal values of some system calls can be higher in malware apps than in goodware apps. This is because malware may invoke some special system call sequences which are not found in goodware apps. These special system call sequences are intended to perform malicious activities. For example, many malware applications invoke sensitive resources such as GPS and camera for performing malicious activities. The main purpose of dup() system call is to perform IPC through shared memory channel. Malware rely on dup() and close() system calls for communicating with sensitive resources. That is, a malware may invoke a malicious system call code consisting of F (dup()) and I (close()) for accessing sensitive resources in the background. In these cases, the signal values of the system calls F (dup()) and I (close()) can become high in malware than in benign applications.

The signal values of those system calls with large difference over goodware and malware applications can be selected as they are more informative in distinguishing malware from goodware. One can select the graph signal values of 'd' such system calls, $1 \le d \le n$, where 'n' is the number of system calls (vertices) in the system call graph as a low dimensional feature vector. The selection of prominent system calls can be done as follows.

Let $\overline{\mathcal{V}}(S_i|malware)$ denotes the mean of the signal values of the system call S_i in a set of m_1 malware application. That is,

$$\overline{\mathcal{V}}(S_i|malware) = \frac{\displaystyle\sum_{malware}\mathcal{V}(S_i)}{m_1} \tag{7.3}$$

Let $\overline{\mathcal{V}}(s_i|goodware)$ denotes the mean of the signal values of the system call s_i in a set of m_2 goodware applications. That is,

$$\overline{\mathcal{V}}(s_i|goodware) = \frac{\sum\limits_{goodware} \mathcal{V}(s_i)}{m_2} \tag{7.4}$$

The variance in the signal values of each system call in the set of malware applications and in the set of goodware applications also need to be computed. A low variance in signal values of a system call indicates that the signal values across application are more centered and consistent in that set of malware or goodware. Let $\overline{D}(s_i)$, $\forall i$ denotes the mean difference. That is,

$$\overline{D}(s_i) = \overline{\mathcal{V}}(s_i|malware) - \overline{\mathcal{V}}(s_i|goodware), \forall i \tag{7.5}$$

Those system calls with mean difference in signal values greater than a threshold can be selected and their signal values can be taken as the features of the application.

Suppose that we have 20 system calls (A,B,...,T) in the system call graph and we obtained the mean, variance and difference of signal values as shown in Table 7.4. In this case, we may proceed with a 20 dimensional feature vector corresponding to the signal values of the 20 system calls (A,B,...,T). However, a dimensionality reduction is possible here. In Table 7.4, we can see that the difference of signal values for the systems calls A,B,E,F,K,N,P and S are above a threshold $t = 0.275$. However, variance is high for the system calls K and P. Hence, we may choose the signal values of A,B,E,F,N and S as a 6 dimensional feature vector instead of a 20 dimensional feature vector and proceed with ML classification.

7.4 Experiments and Analysis

In this section, we illustrate the graph signal based malware detection in a balanced dataset consisting of 1000 malware and goodware applications. The malware applications were downloaded from Drebin/AMD dataset [75] [203] and the goodware applications were from Google play store/Androzoo [28]. In Chapter 4, we selected the malware apps irrespective of their behavior. In this chapter, we selected malware apps which try to access the system resources such as camera, GPS, telephony, etc. in the background. AMD dataset contains some malware apps which evade the detection environment. Here, we have not considered these kinds of evasive malware apps. Also, we have eliminated the duplicate samples from malware and goodware datasets and selected only the unique ones.

7.4.1 Experimental setup

The malware detection system was implemented in an 8 GB Intel core i5 PC with an Android emulator. A dataset of the graph signals $\mathcal{V}_1$ was computed from 1000 malware and goodware samples. In the dataset construction phase, the sample malware

TABLE 7.4 Mean and variances of signals in malware and goodware applications.

SI.No	System Call	$\overline{\mathcal{V}}(\mathcal{S}_i\|$ Malware$)$	Variance	$\overline{\mathcal{V}}(\mathcal{S}_i\|$ Goodware$)$	Variance	$\overline{D}(\mathcal{S}_i)$
1	A	0.639	0.021	0.278	0.038	**0.361**
2	B	0.775	0.021	0.464	0.019	**0.311**
3	C	0.445	0.025	0.309	0.059	0.136
4	D	0.685	0.041	0.570	0.096	0.115
5	E	0.514	0.044	0.200	0.092	**0.314**
6	F	0.452	0.003	0.173	0.090	**0.279**
7	G	0.242	0.076	0.104	0.042	0.138
8	H	0.309	0.073	0.182	0.060	0.127
9	I	0.110	0.037	0.006	0.002	0.104
10	J	0.180	0.057	0.080	0.038	0.100
11	K	0.565	**0.214**	0.275	0.031	**0.290**
12	L	0.755	0.007	0.674	0.025	0.081
13	M	0.655	0.012	0.582	0.038	0.073
14	N	0.813	0.008	0.445	0.029	**0.368**
15	O	0.746	0.003	0.713	0.024	0.033
16	P	0.366	0.026	0.037	**0.217**	**0.329**
17	Q	0.033	0.015	0.011	0.005	0.022
18	R	0.041	0.015	0.023	0.009	0.018
19	S	0.457	0.020	0.049	0.017	**0.408**
20	T	0.007	0.003	0.006	0.002	0.001

and goodware applications were executed in an emulator and the system calls were logged-in using the strace utility [185]. A python code was used to preprocess the relevant system call sequence (file and network management) by eliminating arguments and irrelevant system calls. Then, these preprocessed system calls were modeled as control flow digraphs using the functions in the networkx package. The system call frequencies $F(\mathcal{S}_i) : 1 \leq i \leq n$ and normalized adjacency matrix $\widehat{\mathcal{A}}$ of each application were computed from the control flow graph and the system call sequence. The feature vectors of each application were then computed as the matrix product,

$$\widehat{\mathcal{A}} \times [\frac{F(\mathcal{S}_1)}{\sum\limits_{j=1}^{n} F(\mathcal{S}_j)} \cdots \frac{F(\mathcal{S}_n)}{\sum\limits_{j=1}^{n} F(\mathcal{S}_j)}]^{T}.$$

The python code for graph signal construction is given in the Appendix. This graph signal construction code receives system call sequences of several applications as input and gives the graph signal vectors as the output. In this process, the python code first preprocess the system call sequences by eliminating arguments and irrelevant system calls. Then, it assigns alternative name to system calls instead of their original name for convenience. After that, it extracts system call count values and adjacency matrix from each system call sequence. It then computes the transformed graph signals and save it in the csv file for machine learning classification.

7.4.2   Performance analysis with various ML classifiers

In this section, we discuss about the performance of various machine learning classifiers with graph signals as the feature vectors. The malware detection mechanism involves running any machine learning classifier with graph signals on selected nodes as the feature vector. We have demonstrated the performance with the following five machine learning classifiers:

1. Naive Bayes;
2. SVM (Support Vector Machine);
3. Decision Trees (DT);
4. Random Forest (RF);
5. ANN(Artificial Neural Network).

TABLE 7.5   Performance of various ML classifiers.

Algorithm	TPR	FPR	Precision	Accuracy	F1Score
Naive Bayes	0.81	0.20	0.81	0.81	0.81
SVM	0.83	0.12	0.88	0.86	0.85
ANN	0.93	0.14	0.86	0.89	0.89
Decision Trees (DT)	0.93	0.21	0.82	0.86	0.87
Random Forrest (RF)	0.94	0.13	0.85	0.90	0.90

Here, the signal values of 90% of malware and goodware applications were used for training the machine learning classifiers. The performance of the classifiers were evaluated in the signal values of remaining 10% of malware and goodware applications. The detection performance of the five popular machine learning classifiers with graph signals as features is given in Table 7.5. Here the signal values on all the vertices were taken for the feature vector construction. In Table 7.5, we can see that the signal vector based feature representation gives good performance with SVM, ANN and random forest classifiers. The python codes for the classifiers is given in the Appendix. The file graphsignals.csv is split in the ratio 9:1 for training and testing. That is 90% of samples are used for training and the remaining 10% samples are used for testing. These training and test data sets are given as inputs to various ML algorithms for determining their performance. Here, label_pred is the output (predicted class) of the ML classifier.

7.5   Miscellaneous Operations on Graph Signals

Apart from graph shift operation, there are several other techniques such as graph Fourier transform, and graph filtering which can be used for transforming graph signals to different forms. Some of these transformations may give better signal values and feature vectors. For calculating graph Fourier transform of a signal vector $\mathcal{V}_0$, the generalized eigen vector matrix $\mathcal{F}$ of the adjacency matrix A is computed and multiplied with the graph signals. The eigen vector transformation of the matrix A can be written as:

$$A.\mathcal{F} = \lambda.\mathcal{F},$$

where λ is the eigen value vector. The Fourier transform $\mathcal{V}_1$ of the graph signal $\mathcal{V}_0$ is calculated as

$$\mathcal{V}_1 = \mathcal{F}^{-1}.\mathcal{V}_0.$$

These Fourier transformed signals $\mathcal{V}_1$ can be given as inputs to an ML classifier for malware classification. Also, it is possible for an analyst to design a linear and shift invariant (LSI) graph filter $h(A)$ based on the properties of system call digraphs. The LSI filter matrix $h(A)$ is calculated as:

$$h(A) = \sum_{i=0}^{n} h_i.A^i,$$

where the numerical values h_i, for $i = 1, 2, \ldots, n$ are called the graph filter taps. The filtered signals $\mathcal{V}_3$ are calculated as,

$$\mathcal{V}_2 = h(A).\mathcal{V}_0.$$

7.6   Conclusion

In this chapter, we discussed the graph signal processing based approach for dynamic feature vector construction. The graph signals can be used to detect Android malware with high accuracy while using with various ML classifiers. These graph signals can also be used as additional features for hybrid and ensemble classifiers

Certain malware applications when performing malicious activities, generate system call sequences similar to benign applications[200]. Therefore, if we consider only the system call sequence, these malware apps may get misclassified as goodware apps and vice versa. This can lead to an increase in the false negatives (false positives). Also, in our experiments with new malware applications, we found that multistage malware apps such as camscanner can get undetected if we consider only the system call sequences. Multistage malware apps initially act as a legitimate app and download malicious functionalities from online repositories. The online server communication is done by passing arguments to network management system calls such as sendto() and recvfrom(). Some malware applications inject malicious commands, files, or data as arguments to the system calls such as open(), execve() for performing various malicious activities [141]. The system call execve() is used by goodware as well as malware. Some malware applications request root privileges for exploiting the vulnerabilities in the system utilities for gaining unauthorized access. It uses "/xbin/su" as argument of the system call execve() for checking the root privilege. In all these cases, the malicious behavior may be inferred from the system call arguments rather than the system calls. Hence, one may use the information in system call arguments along with the system call sequences for constructing the feature vectors. It is possible to construct a system call behavioral graph for an application based on the dependencies among system call arguments[134]. However, such kind of argument dependencies are found only in certain kind of applications [134]. Hence, it is not possible to construct these kinds of graphs for all kinds of applications [134]. In order to overcome these limitations, one may consider both the system call argument relationships and control dependency information for building novel graph models.

In Chapter 4 to Chapter 7, we discussed about the efficiency and accuracy of machine learning algorithms for Android malware detection. However, machine learning algorithms need to be frequently retrained in accordance with the emergence of new malware and goodware applications. In order to overcome this limitation, in the next chapter, we will discuss about the existence and extraction of certain short system call subsequences in malware applications. With these subsequences, one can detect malware applications.

8

System Call Pattern-Based Detection

In the previous chapters, we employed machine learning classifiers for malware detection. Machine learning mechanisms usually require good feature representation for accurate classification of the data points. It is a challenge to identify and represent the correct features of the data points. Further machine learning based detection mechanisms can be computationally challenging and may not be suitable for on device real time deployment. Hence, there have been many attempts to detect malware based on various signatures.

According to Artenstein et al. [77], malware applications can utilize the vulnerabilities in the Android IPC to perform various kinds of attacks such as sending SMS, stealing information, etc. A malware app can invoke the services provided by the server processes with the help of sensitive API calls for launching various kinds of attacks. In the kernel level, ioctl () system call is produced for requesting the services provided by the server processes. Then, the kernel verifies the request and creates a shared memory for communication. After the communication, the shared memory is unallocated. Hence, this IPC (Inter Process Communication) between malware application and server processes gets reflected as a short system call pattern in the entire system call sequence. In [190], it is proved that the system call sequence of Android applications contains special patterns which are unique to malware. The system call sequence of Android applications is found to be stationary first-order ergodic Markov chains. As a consequence of this, there exists short system call patterns in the system call sequence of malware applications which contain the malicious information in the entire system call sequence. Even a single pattern can be found in multiple families of malware application. These patterns can be used for efficient signature based malware detection and identification. In this chapter, we discuss how to extract such kind of malicious system call patterns from malware applications and use it for malware detection.

8.1 Extraction of Patterns From System Call Sequences

Most of the malware applications try to access the sensitive resources frequently in the background. It communicates with a sensitive resource silently in the background by generating a system call pattern (contiguous system call sequence between two busy wait system calls) immediately when that resource is available. From this, it is

DOI: 10.1201/9781003121510-8

clear that there exist certain system call patterns (contiguous system call sequence between two busy wait system calls) in a system call sequence which contains relevant information about the malicious behavior. Hence we can identify the malicious behavior in the system call sequence X with such contiguous subsequences in X.

8.1.1 Representing system call sequence as ergodic Markov chain

In the previous chapters, we made detailed discussion about tracing the system call sequence of an application. In the system call sequence of an application, we eliminate the irrelevant system calls and keep only the relevant system calls. The list of relevant system calls $S = \{S_i : i = 1, 2, 3 \ldots, n\}$ is given in Table 8.1. Let, $X = \{X_j : j = 1, 2, \ldots, m\}$ denotes the refined system call sequence after eliminating the irrelevant system calls.

TABLE 8.1 List of relevant system calls.

Alternative Name	A	B	C	D	E	F	G	H
System Call	recvfrom	write	ioctl	read	sendto	dup	writev	pread
Alternative Name	I	J	K	L	M	N	O	P
System Call	close	socket	bind	connect	mkdir	access	chmod	open
Alternative Name	Q	R	S	T	U	V	W	X
System Call	rename	fchown32	unlink	pwrite	umask	lseek	fcntl	recvmsg
Alternative Name	Y	Z	A1	A2	A3	A4		
System Call	sendmsg	epoll	dup2	fchown	readv	chdir		

8.1.2 Computation of information in system call sequence

Let, $T = (t_{ij})$ be the transition probability matrix of the system call sequence. The transition probability t_{ij} from a system call S_i to another system call S_j can be calculated as,

$$t_{ij} = Pr(S_j|S_i) = \frac{n_{ij}}{\sum\limits_{k=1}^{n} n_{ik}}, \forall 1 \leq i, j \leq n \tag{8.1}$$

where n_{ik} is the number of state transitions from a state S_i to the state S_k in X, $\forall 1 \leq k \leq n$.

Since the system call sequence X can be treated as an ergodic Markov chain, $\lim\limits_{k \to \infty} T^k = Q$, where Q is a matrix with identical rows say $\mu = (\mu_1, \mu_2, \ldots, \mu_n)$. That is,

$$\lim_{k \to \infty} T^k = \begin{bmatrix} \mu_1 & \mu_2 & \cdots & \mu_n \\ \cdots & \cdots & \cdots & \cdots \\ \mu_1 & \mu_2 & \cdots & \mu_n \end{bmatrix}$$

μ is known as the *stationary distribution* of the Markov chain X. Entropy $\mathcal{H}(X)$ of a Markov chain is a measure of the information in X. For an ergodic Markov chain X, it is calculated as:

$$\mathcal{H}(X) = -\sum_{i=1}^{n}\sum_{j=1}^{n} \mu_i t_{ij} log(t_{ij}),$$

where $\mathcal{T} = (t_{ij})$ is the transition probability matrix of X and $\mu=(\mu_1, \mu_2,..., \mu_n)$ is the stationary distribution of X. The steps for computing the entropy is given in Algorithm 1. The state transition probability matrix $\mathcal{T} = (t_{ij})$ is given as the input to the Algorithm 1. The algorithm first computes the stationary distribution μ from $\mathcal{T}$ by computing the powers of $\mathcal{T}$ until the rank of the matrix $Q = \mathcal{T}^i$ becomes 1. When the rank becomes 1, all the rows of Q become identical and the first (or any) row of Q gives the stationary distribution μ of X. The entropy rate $\mathcal{H}(X)$ of X is calculated using t_{ij} and μ.

Algorithm 1 Computation of Entropy in a System Call Sequence

Input: $\mathcal{T} = (t_{ij})$
Output: *Entropy*
 1: $i \leftarrow 1$
 2: $Q = (Q_{ij}) \leftarrow \mathcal{T}$
 3: **while** $Rank(Q) \neq 1$ **do**
 4:    $i \leftarrow i + 1$
 5:    $Q =\leftarrow \mathcal{T}^i$
 6: **end while**
 7: $\mu = (\mu_1,\ldots,\mu_n) = (Q_{1,1},\ldots,Q_{1,1})$
 8: $\mathcal{H}(X) \leftarrow -\sum_{i=1}^{n}\sum_{j=1}^{n} \mu_i \times t_{ij} \times log(t_{ij})$

8.1.3   Identification of system call patterns

An application goes through different states during its execution [183]. Those states are:

- Ready : The process is ready to communicate with an available resource;

- Running: The process is communicating with the resources through operating system by means of generating a system call sequence;

- Waiting: The process is waiting for the availability of a resource or user input. An application raises epoll_wait() system call in this state.

A malware application has a tendency to access the system resources during its execution time. An application checks the availability of a resource via polling process . During the polling process, the application checks the status of the file descriptor

of the required resource in order to know the availability of the resource. If the resource is available, the application starts to communicate with the device immediately
resulting in the generation of several system calls. Such system calls generated in between two busy wait calls (epoll_wait) in the system call sequence are extracted and is
stored in the form of system call patterns. It is known that, every application requires
B (write()) and C (ioctl()) system calls for communicating with sensitive resources in
a system [174]. Hence, we may not consider those system call patterns which do not
contain B (write()) and C (ioctl()) system calls.

Let, $X_u^{k_u} = (X_u, \ldots, X_{u+k_u-1})$ be a sequence of system calls between two busy
wait system calls Z. We call $X_u^{k_u}$, a system call pattern. The information distances ϵ_u
between the patterns $X_u^{k_u}$ and the Markov chain of system call sequence X is calculated
as,

$$\epsilon_u = \left| -\frac{1}{k_u} \log \Pr(X_u, \ldots, X_{u+k_u-1}) - \mathcal{H}(X) \right|.$$

The actual behavior of a malware application is concentrated entirely on the *system
call patterns* X_u if the information distance $\epsilon_u \approx 0$.

8.2 System call patterns in Walkinwat trojan

In this section, we explain how to extract the system call patterns from the Walkinwat
trojan and get the best approximation of the system call sequence. It is known that,
Walkinwat trojan performs malicious activities such as sending SMS to all contacts
in a device, stealing information etc. In order to perform this malicious behavior,
Walkinwat malware requires to access the services such as telephony provided by
the server process via IPC based on shared memory mechanism. Hence, the system
call sequence of Walkinwat trojan can contain system call patterns which represent
malicious activities such as accessing system resources like Telephony, and GPS in
the background.

The system call sequence of the Walkinwat trojan is shown in Figure 8.1 and
the system call digraph is shown in Figure 8.2. Consider the system call sequence
Z,A,A,B, ...as a walk in the state transition digraph G. We know that the system
call walk starts from the system call Z. That is, $X_1 = Z$. After context switching, the
application generates a system call sequence {A,A,B,A,C,...,C} and raises the epoll
(Z) system call. Hence, the sequence {Z,A, A,B,A,C,...,C,Z} becomes a closed walk
in the Walkinwat state transition digraph G. This process is repeated several times
during execution of Walkinwat trojan. The transition probability matrix T is given in
Table 8.2.

Now, we know that $\lim_{k \to \infty} T^k = Q$, where Q is a rank 1 matrix with identical rows
say $\mu = (\mu_1, \ldots, \mu_n)$. By raising T to a sufficiently large power k, we can get the matrix
Q and the stationary distribution μ of the Walkinwat trojan as,

$\mu = (0.1900, 0.2954, 0.1841, 0.1626, 0.0031, 0.0045, 0.0023, 0.0019, 0.014, 0.1414).$

Z [AABAACBCBDBDBBBBDBDBBBBDC] Z [DAAA] Z [C] Z [AC] ZZ [AABAACBCBDBDBBBBDBDBBBDC] Z [DAAA] Z [C] Z [AC] ZZ [AABAACBCBDBDBBB
DBDBBBBDC] Z [DAAA] Z [C] Z [AC] ZZ [AABAACBCBDBDBBBBDBDBBBBDC] Z [DAAA] Z [C] Z [AC] ZZ [AABAACBCBDBDBBBBDBDBBBBDC] Z [DAAA]
Z [C] Z [AC] ZZ [AABAACBCBDBDBBBBDBDBBBBDC] Z [DAAA] Z [C] Z [AC] ZZ [AABAACBCBDBDBBBBDBDBBBBDC] Z [DAAA] Z [C] Z [AC] ZZ [AABA
CBCBDBDBBBBDBDBBBBDC] Z [DAAA] Z [C] Z [AC] ZZ [AABAACBCBDBDBBBBDBDBBBBDC] Z [DAAA] Z [C] Z [AC] ZZ [AABAACBCBDBDBBBBDBDBBB
DC] Z [DAAA] Z [C] Z [AC] ZZ [AABAACBCBDBDBBBBDBDBBBBDC] Z [DAAA] Z [C] Z [AC] ZZ [AABAACBCBDBDBBBBDBDBBBBDC] Z [DAAA] Z [C] Z [
AC] ZZ [AABAACBCCBDBDBBBBDBDBBBBDCC] Z [DAAA] Z [C] Z [AC] ZZ [AABAACBCBDBDBBBBDBDBBBBDC] Z [DAAA] Z [C] Z [AC] ZZ [AABAACBCB
DBDBBBBDBDBBBBDC] Z [DAAA] Z [C] Z [AC] ZZ [AABAACBCBDBDBBBBDBDBBBBDC] Z [DAAA] Z [C] Z [AC] ZZ [AABAACBCBDBDBBBBDBDBBBBDC] Z
[DAAA] Z [C] Z [AC] ZZ [AABAACBCBDBDBBBBDBDBBBBDC] Z [DAAA] Z [C] Z [AC] ZZ [AABAACBCBDBDBBBBDBDBBBBDC] Z [DAAA] Z [C] Z [AC] Z
Z [AABAACBCBDBDBBBBDBDBBBBDC] Z [DAAA] Z [C] Z [AC] ZZ [AABAACBCCBDBDBBBBDBDBBBBDC] Z [DAAA] Z [C] Z [AC] ZZ [AABAACBCBDBDBBB
BDBDBBBBDC] Z [DAAA] Z [C] Z [AC] ZZ [AABAACBCBDBDBBBBDBDBBBBDC] Z [DAAA] Z [C] Z [AC] ZZ [AABAACBCBDBDBBBBDBDBBBBDC] Z [DAAA
] Z [C] Z [AC] ZZ [AABAACBCBDBDBBBBDBDBBBBDC] Z [DAAA] Z [C] Z [AC] ZZ [AABAACBCBDBDBBBBDBDBBBBDC] Z [DAAA] Z [C] Z [AC] ZZ [AAB
ACBCBDBDBBBBDBDBBBBDC] Z [DAAA] Z [C] Z [AC] ZZ [AABAACBCBDBDBBBBDBDBBBBDC] Z [DAAA] Z [C] Z [AC] ZZ [AABAACBCBDBDBBBBDBDBBB
BDC] Z [DAAA] Z [C] Z [AC] ZZ [AABAACBCCBDBDBBBBDBDBBBBDCC] Z [DAAA] Z [C] Z [AC] ZZ [AABAACBCCBDBDBBBBDBDBBBBDC] Z [DAAA] Z [
C] Z [AC] ZZ [AABAACBCBDBDBBBBDBDBBBBDC] Z [DAAA] Z [C] Z [AC] ZZ [AABAACBCBDBDBBBBDBDBBBBDC] Z [DAAA] Z [C] Z [AC] ZZ [AABACB
CCBDBDBBBBDBDBBBBDC] Z [DAAA] Z [C] Z [AC] ZZ [AABAACBCBDBDBBBBDBDBBBBDCC] Z [DAAAAACCCBDBDBBBBDBDBBBBDC] Z [AA] Z [A] Z [C]
Z [AC] ZZ [AABAACBCBDBDBBBBDBDBBBBDCC] Z [DAAA] Z [C] Z [AACEA] ZZ [AABAAACCBDBDBBBBDBDBBBBDCBCBDBDBBBBDBDBBBBDC] Z [DAAA
ACBDBDBBBBDBDBBBBDC] Z [AAAEA] Z [A] Z [CCCCCC] Z [AC] Z [AACC] Z [BIBICBI] Z [CCIB] Z [DACCCBDBDBBBBDBDBBBBDC] Z [AA] ZZ [
GCC] Z [AACCCBDBDBBBBDBDBBBBDC] Z [AA] Z [CC] ZZZZZZZ [AACBIBICBICCBI] Z [AC] Z [AA] ZZZ [ACB] Z [DAAA] ZZ [AC] ZZ [AABACB
] Z [D] Z [AABAAC] Z [DAA] ZZ [ICCCCCC] Z [CCIGC] Z [A] Z [CC] ZZZ [CCI] ZZ [C] ZZZ [DCCCC] Z [C] Z [DGDDDDCCCACCCFI] Z [GDCCF
I] Z [CC] Z [AA] Z [AABACCCCC] Z [DAA] ZZ [AACCCCCCFBDIBDBBBDCCCCFBDIBDBBBDCCC] Z [AC] Z [AA] ZZZ [CCCCFI] ZZ [GAAACC
CCCFBDIBDBDBBBDBDBBBDCCBCACCC] Z [DAA] ZZZZ [AACCCCCCCFBDIBDBBBDCCCCFBDIBDBDBBBDBDBBBDC] Z [AA] Z [AAACCCCCC
BDBDBBBDBDBBBDC] Z [AA] ZZ [AACCCBDBDBBBDBDBBBDCC] Z [AA] ZZ [AACCCBDBDBBBDBDBBBDC] Z [AA] Z [AACCCBDBDBBBDBDBBB
DC] Z [AA] Z [AACCBDBDBBBDBDBBBDCC] Z [AA] Z [AACCCBDBDBBBDBDBBBDCC] Z [AA] Z [A] Z [C] ZZ [AC] Z [AAAACBCCFBDIBDBDBBB
DBDBBBDCCBDBDBBBDBDBBBDCC] Z [DAAAEAAAA] Z [AAEEEAEAAACBCCCCFBDIBDBDBBBDBDBBBDCCBDBDBBBDBDBBBDC] Z [DAA] Z [C
] ZZ [AAACBCCBDBDBBBDBDBBBDCCBDBDBBBDBDBBBDC] Z [DAA] Z [CC] Z [BIBICCBI] Z [CCI] Z [AACCBDBDBBBDBDBBBDC] Z [AA] ZZ
[GCC] Z [AACCCBDBDBBBDBDBBBDCC] Z [AA] Z [CCCC] ZZZZZZZ [AACCBIBICCBICCCCBI] Z [AC] ZZ [AAACB] Z [DAAAAC] Z [AA] ZZZ [
A] ZZ [AC] Z [AAACB] Z [DAAA] ZZ [AC] Z [AAACB] Z [DAAA] ZZ [AC] ZZ [AABACB] Z [DAAA] ZZ [ICCCCCC] Z

FIGURE 8.1 System call sequence of Walkinwat Trojan.

TABLE 8.2 State transition probability matrix of Walkinwat Trojan.

	A	B	C	D	E	F	G	H	I	Z
A	0.454	0.098	0.274	0	0.009	0	0	0	0	0.163
B	0.063	0.369	0.063	0.470	0	0	0	0	0.021	0.010
C	0.004	0.268	0.310	0	0.002	0.024	0	0	0.012	0.376
D	0.139	0.659	0.176	0	0	0	0.002	0	0.019	0.002
E	0.714	0	0	0	0.285	0	0	0	0	0
F	0	0.700	0	0	0	0	0	0	0.300	0
G	0.166	0	0.500	0	0	0	0	0.333	0	0
H	0	0	0.400	0	0	0	0	0.600	0	0
I	0	0.480	0.320	0	0	0	0.04	0	0	0.160
Z	0.414	0.006	0.147	0.167	0	0	0.009	0	0.006	0.247

The entropy of the system call sequence of Walkinwat trojan can be calculated using $\mathcal{T}$ and μ as,

$$\mathcal{H}(X) = -\sum_{i=1}^{n} \sum_{j=1}^{n} \mu_i t_{ij} log(t_{ij}) = 1.7512.$$

The system call patterns generated by the Walkinwat trojan and their information distance ϵ are given in Table 8.3. The consecutive occurrence of same system calls in a system call pattern are represented using power notation. For example, a system call

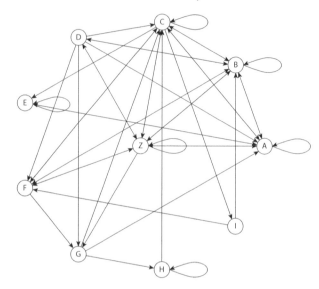

FIGURE 8.2 System call digraph of Walkinwat malware application.

TABLE 8.3 System call patterns and information distance values.

ID	System Call Pattern	ϵ
Pattern 1	$A^2C^7FBDIBDB^3DC^4FBDIBDBDB^3DBDB^3DC$	0.05
Pattern 2	$A^2E^3AEA^3CBC^4FBDIBDBDB^3DBDB^3DC^2BDBDBDB^3DBDB^3DC$	0.07
Pattern 3	$GA^3C^5FBDIBDBDB^3DBDB^3DC^2BCAC^3$	0.18
Pattern 4	$A^2CBIBICBIC^2BI$	1.05

pattern AACBIBICBICCBI is represented in power notation as $A^2CBIBICBIC^2BI$ for convenience. Here we can select the system call pattern with least information distance (Pattern 1) as the *malicious system call pattern* of the Walkinwat trojan.

8.3 Malware Detection and Classification Based on System Call Patterns

System call pattern based malware detection and classification involves comparing the system call pattern of the test applications with the system call patterns of known malware. So we need to first build a dataset of system call patterns of known malware. The system calls generated by various malware samples have to be refined and the system call patterns are to be extracted out as described in the Section 8.1. Each pattern is then compared with every *malicious system call pattern* stored in the database (created earlier). The comparison is carried out using the Jaro-Winkler similarity metric [155]. The system calls in a system call pattern are repetitive in nature. Usually, a few initial system calls in similar system call patterns are equal. Hence, similar system call patterns have a common prefix. Therefore, Jaro-Winkler similarity metric can be used to compare the system call patterns as Jaro-Winkler considers common prefix for comparison.

Jaro-Winkler similarity metric is a string metric for measuring the edit distance between two sequences. Higher the Jaro-Winkler distance for two string, the more similar the strings are. The score is normalized such that 1 corresponds to exact match and 0 corresponds to perfect dissimilarity. If the match score is greater than a threshold $T \in [0, 1]$, the application can be flagged off as malicious and the process can be terminated. Otherwise, the similarity comparison is to be carried out with other system call patterns in the test application. If the Jaro-Winkler similarity score is less than or equal to the threshold T in all cases, the application can be declared as a goodware. The malware detection steps are given in Algorithm 2.

Let P_1 and P_2 be the two system call patterns. Then, the Jaro score J of P_1 and P_2 is given by:

$$J = \frac{1}{3}\left(\frac{m}{|P_1|} + \frac{m}{|P_2|} + \frac{(m-t)}{m}\right), \tag{8.2}$$

where m is the number of matched system calls, t is the number of transpositions and $|P_1|, |P_1|$ are the length of these patterns. The Jaro-Winkler score J_w is calculated as:

$$J_w = J + (lp(1 - J)), \tag{8.3}$$

where l is the length of common prefix and p is a scaling constant with value equal to 0.1. Two characters from P_1 and P_2, are considered matching only if they are the same and not farther than

$$\lfloor \frac{\max(|P_1|, |P_2|)}{2} \rfloor - 1.$$

Each character of P_1 is compared with all its matching characters in P_2. The number of matching (but different sequence order) characters divided by 2 defines the number of transpositions.

The malware detection mechanism using system call pattern matching with Jaro-Winkler score is given in Algorithm 2. The set of *malicious patterns* $MP = \{MP_1, \ldots, MP_l\}$ obtained in the training phase from various malware families, the set of system call patterns $P = \{P_u = (P_{u,v} : v = 1, 2, \ldots, k_u) : u = 1, 2, \ldots, k\}$ obtained by running the test application (Section 8.3) and a threshold value T are given as input to the Algorithm 2. *State* variable holds the value 0 to denote a goodware and 1 to denote a malware. In Algorithm 2, $|MP_i|$ denotes the length of the ith *malicious pattern*, $|P_u|$ denotes the length of the uth system call pattern in the test application, $m_{i,u}$ denotes the number matching system calls, $tr_{i,u}$ denotes the number of transpositions and $l_{i,u}$ denotes the common prefixes between the system call patterns MP_i and P_u. First, the Jaro score J and then the Jaro-Winkler score are calculated for each pair of *malicious pattern* $MP_i \in MP$ and system call pattern $P_u \in P$. If the Jaro-Winkler similarity score exceeds a predetermined threshold value T for any such pair, then the application is flagged off as a malware, otherwise, the application is flagged off as a goodware.

Algorithm 2 Malware Detection (System Call Pattern Matching) Algorithm

---

Input: $MP = \{MP_1, \ldots, MP_l\}$, $P = \{P_u : u = 1, 2, \ldots, k\}$, T
Output: $State$

1: $i \leftarrow 1$
2: $j \leftarrow 1$
3: $State \leftarrow 0$
4: **for** $i \in \{1, 2, \ldots, l\}$ **do**
5:     **for** $u \in \{1, 2, \ldots, k\}$ **do**
6:         $J \leftarrow \frac{1}{3}\left(\frac{m_{i,u}}{|MP_i|} + \frac{m_{i,u}}{|P_u|} + \frac{(m_{i,u} - tr_{i,u})}{m_{i,u}}\right)$
7:         $J_w \leftarrow J + 0.1 \times l_{i,u}(1 - J)$
8:         **if** $J_w \geq T$ **then**
9:             $State \leftarrow 1$
10:        **end if**
11:    **end for**
12: **end for**

8.4   Experiments and Analysis

In this section, let us examine the performance of the *malicious system call pattern* based malware detection mechanism. We consider an initial dataset of 2000 samples comprising of goodware as well as malware applications. Goodware applications are downloaded from the Google play store [28] and malware applications are downloaded from the Drebin and the AMD datasets [75] [203]. Further, we downloaded Judy infected autoclicking adware applications and ransomware from external repositories [113] [114] [115]. The tested malware applications include spyware privilege escalators, SMS senders, adware and ransomware, etc. Spyware are applications can perform many malicious activities like stealing information such as IMEI code, monitoring the location, listening phone calls, sniffing the surroundings by controlling camera etc. of a compromised device. Privilege escalators are malware applications which make use of the vulnerabilities in a device for gaining access to its protected resources without owner's knowledge. SMS senders can send SMS messages to premium rate numbers stealthily in the background. Spyware, privilege escalators and SMS senders can be categorized as Trojans. These malware can perform malicious activities without any user intent. Adware loads advertisements from a server for generating revenue to the attacker. Certain kinds of adware can generate auto clicks on these advertisements. Ransomware are of two types: Crypto ransomware and locker ransomware. Crypto ransomware encrypts the files in a device and demands a ransom for decrypting it. Locker ransomware blocks the device from usage by locking the screen until the user pays a ransom to the attacker. The tested goodware apps includes gaming apps, social apps etc.

The methodology was implemented in a 16 GB Intel core i5 PC with an Android emulator having ARM v8 architecture. This emulator can simulate the hardware resources in a real Android device. In the pattern extraction phase, the sample malware applications are executed in an emulator by injecting 1000 random events using monkey runner tool. Then, the generated system calls are logged using strace utility.

The malicious system call pattern of three categories of malware applications (Trojans, malvertising applications, and ransomware) were extracted out. Trojans are malware applications which tend to access the system resources in the background. The trojans considered for pattern extraction are Android.Tapsnake (Trojan Spy), Android.wakinwat (Trojan SMS) and Droidkungfu application (privilege escalator). Malvertising applications tend to generate many advertisements for revenue to a third party. In this category, we considered the Judy malware for pattern extraction. Ransomeware encrypt the files or lock the device for a ransom. We have analyzed Simplocker crypto ransomware and VideoPlayer locker ransomware. The system call patterns generated by these malware applications are given in Table 8.4. In Table 8.4, we can see that there exist a longest common subsequence (*LCS*) among the system call patterns of a particular category of malware. Hence, we can identify the category from the system call patterns generated by an application. The system call patterns generated by goodware applications are given in Table 8.5. Unlike in the case of malware, we may not be able to find longest common subsequences (*LCS*) in goodware applications. The system call sizes of malware samples used in pattern extraction is given in Table 8.6. The python code for extracting the system call pattern of an application is given in the Appendix.

TABLE 8.4   System call patterns of various malware.

Pattern ID	Application Category	Application Type	Application Name	*System Call Pattern*	*LCS*
Pattern 1	Trojan	Trojan Spy	Android.TapSnake	AABCCFBDIBDBDBBBDBDBBBDC	
Pattern 2	Trojan	Trojan SMS	Android.Wakinwat	AACCCCCCCFBDIBDBBBDCCCC FBDIBDBDBBBDBDBBBDC	FBDIBD
Pattern 3	Trojan	Privilage Escalator	DroidKungFu	ACCCCCCFBDIBDBBBDC	
Pattern 4	Malvertising Application	Autoclicking Adware	Judy	GCGCCCCGCCCCCCCCGBP DDDDDDDDDDDDDDDDDDDDDD DDDDDDDDDDDDDDDDDDDDDD DDDDDDDDDDDDDDDDDDDDDD DDDDDDDDDDDDDDDDDDDDDD DDDDDDDDDDDDDDDDDDDDDD DDDDDDDDDDDDDDDDDDDDDD DDDDDDDDDDDDDDDDDDDDDD DDDDDDDDDDDDDDDDDDDDDD DDDDDDDDDDDDDDDDDDDDDD DDDDDDDDDDDDDDDDDDDDDD DDDDDDDDDDDDDDDDDDDDDD DDDDDDDDDDDDDDDDDDDDDD DDDDDDDDDDDDDDDDDDDDDD DDDDDDDDDDDDDDDDDDDDDD DDDDDDDDDDDDDDDDDDDDDD DDDDDDDDDDDDDDI	
Pattern 5	Ransomware	Crypto Ransomware	Simplocker	DAAAEAABACBDBDCCCCB	
Pattern 6	Ransomware	Locker Ransomware	Videoplayer	DAAABACBDBDBDGBDBDBDBDBD DBDBDBDBDBDBDBDBDGBDBDBD BDBDBDBDBDBDBDBDBDBDBDBDBDBD DBDBDBDBDBDBDBDBDGBDBDBDCCCB	BACBDBD

In the testing phase (malware detection by comparison with known system call patterns), 2000 samples of goodware and malware applications were executed in an

TABLE 8.5   System call Patterns of various goodware.

Application Category	Pattern ID	System Call Pattern
Calculo IMC App	Pattern 1	AABCBDDBBDDBDDCCBDDCCBDDBDDBDDCCBDDC
Calculator App (System)	Pattern 2	AACBDDBDDBDDCCCB
Superb Booster App	Pattern 3	DAAAACBBBBBDBDCCB

TABLE 8.6   System call size of malware samples used in pattern extraction.

Category	Application	Number of Events	Size of the Trace
Trojan	Android.TapSnake	1000	4703
Trojan	Android.WakinWat	1000	2182
Trojan	DroidKungFu	1000	1814
Malvertising App.	Judy	1000	38007
Ransomware	Simplocker	1000	1214
Ransomware	Videoplayer	1000	8472

emulator and the system calls were collected. These system call logs were segmented and refined. Then, the Jaro-Winkler similarity scores of system call patterns in each application with the known system call patterns of malware from the training phase (see Table 8.4) were computed.

It was found that the Jaro-Winkler similarity score of any application was bounded below by 0.66. This is because every application tend to generate patterns with single system calls (see Figure 8.1) such as read(), recvfrom(), and ioctl() for performing simple operations like reading datagram from the device drivers. The system call patterns also contain at least one ioctl(), read(), and recvfrom() system call. Therefore, system call patterns with single system calls such as ioctl(), read() or recvfrom() have Jaro Winkler similarity greater than or equal to 0.66. This can be proved as follows.

Let P_1 be a pattern in an application with a single system call such as read() or recvfrom() or ioctl() and P_ϵ be a syatem call pattern of length l_ϵ. Then the Jaro score between P_1 and P_ϵ is

$$J = \frac{1}{3}\left(\frac{m}{|P_1|} + \frac{m}{|P_\epsilon|} + \frac{(m-t)}{m}\right)$$
$$= \frac{1}{3}\left(\frac{1}{1} + \frac{1}{l_\epsilon} + \frac{(1-0)}{1}\right) \geq \frac{2}{3} = 0.66.$$

(8.4)

Note that $m = 1$ and $t = 0$, since there is only one system call in P_1. Now the Jaro-Winkler score J_w is,

$$J_w = J + (lp(1-J)) \geq 0.66$$

(8.5)

Thus, the maximum Jaro Winkler similarity score of any application has a minimum value of 0.66.

The malware detection involves computing the Jaro-Winkler similarity score between the system call pattern of the test application with the system call patterns of known malware applications in the database. If the Jaro Winkler similarity score

TABLE 8.7 Performance results.

Dataset	TPR	TNR	FPR	FNR	PPV	Accuracy	F1_Score
Drebin +AMD+ External Repositories	0.940	0.961	0.039	0.060	0.95	0.949	0.95
Drebin	0.935	0.961	0.039	0.065	0.959	0.948	0.947
AMD Dataset	0.940	0.961	0.039	0.06	0.96	0.95	0.955
External Repositories	1	1	1	1	1	1	1

is above a pre-determined threshold, then the test application will be considered as a malware. The FPR and FNR values against various threshold values (with system call patterns of 6 malware) are shown in Figure 8.3. From Figure 8.3, we can conclude that $T = 0.85$ serves as a good threshold to separate malware and goodware applications. The detection rates of the proposed approach in various databases are given in Table 8.7. From Table 8.7, we can conclude that the system call pattern based mechanism can effectively detect malicious applications with an accuracy and F1 score of around 0.95. The detection rates of different malware families are given in Table 8.8.

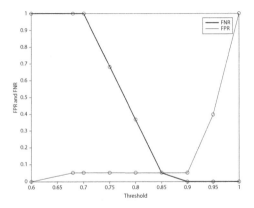

FIGURE 8.3 False positive and negative rates against different threshold values.

The python code for the Jaro-winkler similarity matching for the system call patterns is given in the Appendix.

8.5 Conclusion

In this chapter, we discussed a malware detection mechanism which makes use of the presence of system call patterns in the system call sequence of the application. Initially, the system call patterns of known Android malware are required to be identified and the distinct ones have to be stored as malware signatures. A single system call pattern can represent many malware families. Malware detection works by checking the presence of any of the known system call patterns in the system call sequence

TABLE 8.8   Rate of detection across malware families.

Malware Family	Tested Samples	Category	Rate of Detection
Adrd	20	Trojan	1
Andrup	45	Trojan	0.92
Base Bridge	20	Trojan	0.80
DrodKungFu	147	Trojan	0.91
GoldDream	37	Trojan	0.92
AndroRAT	45	Trojan	0.97
Boxer	44	Trojan	0.88
MobileTX	17	Trojan	0.64
Rumms	50	Trojan	1
Lottor	100	Trojan	0.90
Mseg	128	Trojan	0.92
Penetho	18	Trojan	1
Vidro	23	Trojan	1
Mercor	178	Trojan	1
MMarketPay	9	Trojan	0.77
Lnk	5	Trojan	1
FakeDoc	21	Trojan	1
FakeAV	5	Trojan	1
FakeInstall	15	Trojan	1
FakePlayer	16	Trojan	1
FakeTimer	10	Trojan	1
Simplocker	19	Ransomware	1
VideoPlayer	18	Ransomware	1
Judy	11	Malvertising Application	1

of the application. This malware detection approach can overcome the limitations of many existing machine learning based detection mechanisms.

Reinforcement learning techniques can be used to identify the malicious behavior even if there is a change in system call sequence. That is, one can use the temporal difference algorithm to learn the properties of known malicious system call patterns and identify the malicious probability value of unknown system call patterns [192] [215]. Further, using the system call arguments along with the system calls may give better characterizations of malware families.

9

Conclusions and Future Directions

In this chapter, we discuss about the trends in emerging malware, effectiveness of the detection mechanisms described in Chapter 4 to Chapter 8 and the possible research directions to overcome the limitations of these malware detection mechanisms. The suggested extensions of the detection mechanisms can detect and prevent most of the malware attacks.

In Chapter 4, we showed that malicious behavior can be accurately detected by combining API calls, permissions and system calls. However, an attacker can evade the static malware classifiers by using adversarial techniques. In such cases, static and hybrid analysis mechanisms are not very useful. However, it is very hard for an attacker to alter the system call sequence of an application without modifying the program semantics. Hence, it is possible to identify the malicious behavior from the system call sequence itself. However, in current settings the system call analysis is not very accurate because of the limited code coverage problem . Moreover, it is restricted to the malware which try to access sensitive resources in the background. In order to overcome these limitations, we need to employ mechanisms to cover all the parts of an application source code during its execution time. Further, we need to develop techniques to identify the malicious behavior in unseen system call sequences.

9.1   Recent Malware Attacks

During the COVID-19 outbreak, cybercriminals used COVID-19 tracker applications as a bait to carry out various malicious activities. The spread of coronavirus helped the attackers to launch malicious applications that disguise as covid trackers. Most of these malware appeared in the form of trojans, spyware and ransomware. These malware use anti-detection techniques to prevent them from being detected. Further, the spyware and ransomware attacks dramatically increased during this time leading to huge financial losses and serious security breaches.

Pegasus is a spyware made by the Israeli company called NSO Group[46]. Pegasus can extract the personal details of the user and can access the microphone and camera of the device without user's permission. It can also access the device's location, read text messages and can track calls. The earlier version of Pegasus was installed on smartphones by spear-phishing. However, the latest version of Pegasus spreads through a 'zero click' attack in which the malware enter into the device without clicking a malicious link or user interaction. This malware has the ability to get installed with a missed call on WhatsApp, and can remove the record of the missed call after installation. Another installation technique is by sending a message to a user's phone without any notification. In Android operating system, pegasus uses a rooting method called framaroot to gain access into the device. This method of rooting is subtle and hence the user thinks that the device is not infected.

Due to the popularity of bitcoins, cybercriminals are now exploiting Android applications to spread cryptocurrency-mining malware[61]. These cryptocurrency-mining malware are distributed through Google Play Store and other third party app stores disguised as, gaming app, streaming apps, and VPNs. Most of the disguised applications appear in the form of applications that are used to watch football matches. A common method that is used by the attackers is to conceal a Coinhive JavaScript miner in the Android application. When the user watches the broadcast, the application opens an HTML file with the embedded JavaScript miner. This javascript converts the user's CPU power into a tool for mining cryptocurrency. Crypto-currency mining malware also appeared in the form of adware. Recently, a crypto-currency mining malware family called Trojan.AndroidOS.Coinge.j emerged, that installs itself as a porn app or as a system app. This malware was used by the attackers to monitor the device's battery and temperature to regulate mining activities. The cryptominers are a real threat since they can adversely affect the device performance and can also trigger network based attacks.

9.2   Identifying Exploitation Attacks

Wagner et al. [200] had shown that, certain malware applications when performing malicious activities, generate system call sequences similar to goodware applications. If we consider only the system call sequence, these malware apps may get wrongly classified as goodware apps (and vice versa) which can lead to an increase in the false negatives (false positives). In these cases, the malicious behavior may be identified from the system call arguments along with the system call sequences. For instance, a malware app can perform malicious activities by exploiting the buffer overflow vulnerabilities in the system utilities or programs [141, 90, 84]. That is, a malware app can try to inject malicious commands or files or data as arguments to certain kinds of system calls such as open(), execve() for performing malicious activities [141]. The system call execve() is used by goodware as well as malware. A malware requests root privileges for exploiting the vulnerabilities in the system utilities for gaining unauthorized access. A malware application uses "/xbin/su" as argument of the system call execve() for checking the root privilege. Thus, in these cases, the malicious behavior may be inferred from the system call arguments rather than the system calls.

Experiments with new malware apps have shown that multistage malware apps such as camscanner can escape many detection mechanisms. Multistage malware apps initially act as legitimate apps and later download malicious codes from online servers. The online server communication is done by passing arguments to network management system calls such as sendto() and recvfrom(). In this case, the detection mechanisms discussed in this book may not work. In [134], Kolbitch et al. suggested a mechanism to construct a system call behavioral graph for an application based on the dependencies among system call arguments. However, such kinds of argument dependencies are found only in certain kinds of applications [134]. Hence, it is not possible to construct these kinds of graphs for all kinds of applications [134]. In order to overcome the limitations of not using the system call arguments, one may consider both the system call argument relationships and control dependency information for building graph models.

9.3   Mitigating Emulator Evasion and Code Coverage Problem

Many malware detection mechanisms including the ones discussed in this book, rely on Android emulators to trace out the system calls produced by an application. However, certain malware applications can bypass the emulator based detection mechanisms by using techniques such as verifying the presence of motion sensor, checking for the null value in IMEI code, etc.[166]. This can be resolved by employing techniques such as modifying the emulator, realistic simulation of sensor events, acute binary translation and hardware assisted virtualization [166],[166].

Dynamic malware detection systems detects a malware only if the malware exhibits malicious behavior at least once in an execution. In this book, we used monkey runner tool to inject pseudo random events such as click event, touch event, etc. during its runtime. Monkey runner relies on random exploration strategy to generate input test cases. Certain malware applications such as Xhelper performs malicious activities only when user presses a particular key or when a particular event occurs. Further, it is possible for a malware application to extend its functionality by downloading malicious codes from online repositories during its runtime [167]. This behavior is known as dynamic loading . In such cases, the malware app can update itself or load some malicious code from online sources after few minutes of execution or when a particular event occurs. In those cases, a malware application might get misjudged as benign. In order to overcome these limitations, we can employ mechanisms to cover all parts of the source code of an application during execution. For this, we can use forward execution based on the application control flow graph while collecting the system call sequence [221]. That is, we initially build a control flow graph based on the source code of the application and then inject user or system events to explore all the possible paths in that control flow graph. This collected system call sequence will contain all the possible information related to the application and can be used to identify the malicious behavior from it.

9.4   Resilience to the Change in System Call Sequence

We can employ reinforcement learning techniques to identify the malicious behavior even if there is a change in the system call sequence of a malware. That is, we can use temporal difference algorithm to learn the properties of known malicious system call patterns and identify the malicious probability value of unknown system call patterns [192] [215]. Also, we can consider the system call arguments along with the system call sequence for constructing stochastic models.

9.5   Collusion Attack

In this book, we have considered only the malicious behavior of a single application. We have not considered collusion attacks from multiple applications. In a collusion attack, the malicious code is fragmented into several parts and distributed across multiple apps [150]. These apps communicate together via covert channels for performing malicious activities. For example, an app 'A' steals the information and sends it to an app 'B'. The app 'B' sends that information to the attacker. In such a situation, the background communication with other processes may get reflected in the system call sequences of all the colluding apps in the device. We haven't considered these kinds of malicious attack (collusion) jointly conducted by more than one application (malicious behavior in multiple app scenarios). We need to find some novel mechanisms to detect such kind of malicious behavior in multiple app scenario by correlating the system call sequences generated by all the applications in a device.

Appendix

Source Codes of Chapter 3

The following shell script can be used to extract opcodes from Android application.

```
for file in ./*.apk; do apktool d -f $file;done
for dir in $(ls -d */);
do
mkdir ${dir}smali/opcodes
FILES=$(find ${dir}smali -type f -name '*.smali')
for file in $FILES;
do
  grep -v [0*.#:}{\"] $file | sed '/^$/d' |
  sed's/ //g' | cut -d","
  -f1 | sort -u  >> ./${dir}smali/opcodes/opcodes.txt
done
done
for dir in $(ls -d */);
do
   sed "s/v[0-9]//" ./${dir}smali/opcodes/opcodes.txt >
    ./${dir}smali/opcodes/opcodesv1.txt
   sed "s/p[0-9]//" ./${dir}smali/opcodes/opcodesv1.txt >
   ./${dir}smali/opcodes/opcodesv2.txt
   sed "s/[0-9]$//" ./${dir}smali/opcodes/opcodesv2.txt >
   ./${dir}smali/opcodes/opcodesv3.txt
done
```

Source Codes of Chapter 4

The source code for extracting API, permission and system call based features are given below. Here, Androguard tool [98] is used to extract API calls and permissions from the applications to construct csv files.

```
from androguard.misc import AnalyzeAPK
```

```
import os
import time
import array as arr
import numpy
from numpy import zeros
path = "C:/MalwareDataset"
path1 = "C:/GoodwareDataset"
file1=open("permission.csv","w")
file2=open("apicall.csv","w")
dirs = os.listdir(path)
dirs1=os.listdir(path1)
a=[]
a1=[]
permissions=[]
apicalls=[]
l=['android.permission.WRITE_SETTINGS',
'android.permission.ACCESS_NETWORK_STATE',
'android.permission.READ_EXTERNAL_STORAGE',
'android.permission.ACCESS_MOCK_LOCATION',
'android.permission.USE_CREDENTIALS',
'android.permission.HARDWARE_TEST',
'android.permission.GET_ACCOUNTS',
'android.permission.SEND_SMS',
'android.permission.VIBRATE',
'android.permission.READ_HISTORY_BOOKMARKS',
'android.permission.ACCESS_COARSE_LOCATION',
'android.permission.EXPAND_STATUS_BAR',
'android.permission.UNINSTALL_SHORTCUT',
'android.permission.READ_LOGS',
'android.permission.ACCESS_GPS',
'android.permission.CHANGE_NETWORK_STATE',
'android.permission.FACTORY_TEST',
'android.permission.INSTALL_SHORTCUT',
'android.permission.CHANGE_WIFI_STATE',
'android.permission.SYSTEM_ALERT_WINDOW',
'android.permission.WRITE_SMS',
'android.permission.KILL_BACKGROUND_PROCESSES',
'android.permission.MODIFY_PHONE_STATE',
'android.permission.DEVICE_POWER',
'android.permission.LOCATION',
'android.permission.RECEIVE_BOOT_COMPLETED',
'android.permission.WAKE_LOCK',
'android.permission.WRITE_APN_SETTINGS',
'android.permission.ACCESS_COARSE_UPDATES',
'android.permission.WRITE_HISTORY_BOOKMARKS',
```

```
'android.permission.WRITE_CONTACTS',
'android.permission.PROCESS_OUTGOING_CALLS',
'android.permission.SET_WALLPAPER',
'android.permission.CALL_PHONE',
'android.permission.ACCESS_LOCATION_EXTRA_COMMANDS',
'android.permission.INTERNET',
'android.permission.ACCESS_FINE_LOCATION',
'android.permission.READ_SMS',
'android.permission.RECEIVE_SMS',
'android.permission.BROADCAST_STICKY',
'android.permission.GET_TASKS',
'android.permission.WRITE_EXTERNAL_STORAGE',
'android.permission.RESTART_PACKAGES',
'android.permission.MOUNT_UNMOUNT_FILESYSTEMS',
'android.permission.REBOOT',
'android.permission.INSTALL_PACKAGES',
'android.permission.ACCESS_WIFI_STATE',
'android.permission.DISABLE_KEYGUARD',
'android.permission.READ_CONTACTS',
'android.permission.MODIFY_AUDIO_SETTINGS',
'android.permission.READ_PHONE_STATE',
'android.permission.ADD_SYSTEM_SERVICE',
'android.permission.ACCESS_LOCATION']
file1.write(str(l))
l1=['exec','getDeviceId','getLatitude',
'abortBroadcast','takePicture',
'getLongitude','getWifiState','getDeviceInfo',
'createFromPdu','getSimOperatorName',
'getPackageInfo','requestFocus','getAccountName',
'getIMEI','getCertStatus','getAppPackageName',
'getCellLocation','setSerialNumber','sendSMS',
'getCredential','getSessions','getCookies',
'getSignalLevel','getMessage','getDisplayMessageBody',
'getClassLoader','loadClass','getMethod',
'getDisplayOriginatingAddress','getInputStream',
'getOutputStream','killProcess','getLine1Number',
'getNetworkOperator','getNetworkType',
'getSimSerialNumber','getSubscriberId',
'getLastKnownLocation','isProviderEnabled',
'sendTextMessage']
file2.write(str(l1))
for f in dirs1:
    location="C:/GoodwareDataset"+f
    #print(location)
    #file2=open(location,'r')
```

```
apicall=[]
a=[]
a1=[]
permissions=[]
try:
    app, list_of_dex, dx = AnalyzeAPK(location)
    permissions=app.get_permissions()
    for method in dx.get_methods():
        apicall.append(method.name)

    for i in range(0,len(l)):
        if l[i] in permissions:
            a.append(1)
        else:
            a.append(0)
    for i in range(0,len(l1)):
        if l1[i] in apicall:
            a1.append(1)
        else:
            a1.append(0)
    file1.write('\n')
    file1.write(str(a))
    file1.write(",0")
    file2.write('\n')
    file2.write(str(a1))
    file2.write(",0")
except:
    print('Exception')
for f in dirs:
    location="C:/MalwareDataset/"+f
    apicall=[]
    a=[]
    a1=[]
    permissions=[]
    try:
        app, list_of_dex, dx = AnalyzeAPK(location)
        permissions=app.get_permissions()
        for method in dx.get_methods():
            #print(apicall)
            apicall.append(method.name)

        for i in range(0,len(l)):
            if l[i] in permissions:
                a.append(1)
            else:
```

```
                    a.append(0)
            for i in range(0,len(l1)):
                if l1[i] in apicall:
                    a1.append(1)
                else:
                    a1.append(0)
            file1.write('\n')
            file1.write(str(a))
            file1.write(",1")
            file2.write('\n')
            file2.write(str(a1))
            file2.write(",1")
        except:
            print('Exception')
file1.close()
file2.close()
path = "C:\\MalwareDataset"
path1= :C:\\GoodwareDataset"
files1=os.listdir(path1)
File3 = open('C:\\syscalls.csv','w')
files=os.listdir(path)
z=[]
z57=[]
File3.write("A,B,C,D,E,F,G,H,I,J,K,
L,M,N,O,P,Q,R,S,T,U,V,W,X,Y,Z,Application")
for Files in files:
    # Reading system call sequence
    with open(path+"/"+Files) as f:
        for line in f:
            z=line.split("(")[0]
            z57=z57+z
    fv=[z57.count('A'),z57.count('B'),z57.count('C'),
    z57.count('D'),z57.count('E'),z57.count('F'),
    z57.count('G'),z57.count('H'),z57.count('I'),
    z57.count('J'),z57.count('K'),z57.count('L'),
    z57.count('M'),z57.count('N'),z57.count('O'),
    z57.count('P'),z57.count('Q'),z57.count('R'),
    z57.count('S'),z57.count('T'),z57.count('U'),
    z57.count('V'),z57.count('W'),z57.count('X'),
    z57.count('Y'),z57.count('Z')]
    File3.write("\n")
    File3.write(str(fv))
    File3.write(",1")
    for Files in files1:
    # Reading system call sequence
```

```
with open(path+"/"+Files) as f:
    for line in f:
        z=line.split("(")[0]
        z57=z57+z
fv=[z57.count('A'),z57.count('B'),z57.count('C'),
z57.count('D'),z57.count('E'),z57.count('F'),
z57.count('G'),z57.count('H'),z57.count('I'),
z57.count('J'),z57.count('K'),z57.count('L'),
z57.count('M'),z57.count('N'),z57.count('O'),
z57.count('P'),z57.count('Q'),z57.count('R'),
z57.count('S'),z57.count('T'),z57.count('U'),
z57.count('V'),z57.count('W'),z57.count('X'),
z57.count('Y'),z57.count('Z')]
File3.write("\n")
File3.write(str(fv))
File3.write(",0")
```

Source Codes of Chapter 5

The code for representing a system call sequence as digraph and extracting the centrality measures such as eignvector centality, betweeness centrality and closeness centrality is given below.

```
import csv
import networkx as nx
import math
from sklearn import preprocessing
import numpy
import os, sys
import re
import numpy as np
PQ=[]
path = "C:\\Dataset"
myFile1 = open('C:\\GraphSignals\\eigencent.csv',
'w')
myFile2 = open('C:\\GraphSignals\\betcent.csv',
'w')
myFile3 = open('C:\\GraphSignals\\Closecent.csv',
'w')
files=os.listdir(path)
l=[]
p=[]
i=1
```

```
for Files in files:
    l1=[]
    # Reading system call sequence
    with open(path+"/"+Files) as f:
        for line in f:
            z=line.split("(")[0]
            z1=z1+z
    z12345=[]
    d=[]
    e=[]
    b=[]
    c=[]
    #Create adjacency matrix
    def rank(c):
        return ord(c) - ord('A')
    T = [rank(c) for c in z9]
    M = [[0]*26 for _ in range(26)]
    for (i,j) in zip(T,T[1:]):
        M[i][j] += 1
    for row in M:
        n = sum(row)
        if n > 0:
            row[:] = [f/sum(row) for f in row]
    #Graph construction from adjacency matrix
    G=nx.DiGraph()
    G.add_node("A")
    G.add_node("B")
    G.add_node("C")
    G.add_node("D")
    G.add_node("E")
    G.add_node("F")
    G.add_node("G")
    G.add_node("H")
    G.add_node("I")
    G.add_node("J")
    G.add_node("K")
    G.add_node("L")
    G.add_node("M")
    G.add_node("N")
    G.add_node("O")
    G.add_node("P")
    G.add_node("Q")
    G.add_node("R")
    G.add_node("S")
    G.add_node("T")
```

```
G.add_node("U")
G.add_node("V")
G.add_node("W")
G.add_node("X")
G.add_node("Y")
G.add_node("Z")
n=[]
n1=[]
for i in range(0,len(z57)-1):
    G.add_edge(z57[i],z57[i+1])
for i in range(65,91):
    for j in G.neighbors(chr(i)):
        n.append(j)
    n1.append(n)
# Extraction of centrality measures
eigen=nx.eigenvector_centrality_numpy(G)
betweenness=nx.betweenness_centrality
(G,normalized=False)
closeness=nx.closeness_centrality(G)
   for i in range(65,91):
    try:
        eig1=eigen.get(chr(i))
        e.append(round(eig1,4))
    except KeyError:
        e.append(0)
for i in range(65,91):
    try:
        deg1=degree.get(chr(i))
        d.append(deg1)
    except KeyError:
        d.append(0)
for i in range(65,91):
    try:
        between1=betweenness.get(chr(i))
        b.append(round(between1,4))
    except KeyError:
        b.append(0)
for i in range(65,91):
    try:
        closeness1=closeness.get(chr(i))
        c.append(round(closeness1,4))
    except KeyError:
        c.append(0)
# Logging centralities as csv files
m=[]
```

```
m1=[]
myFile1.write("\n")
myFile2.write("\n")
myFile3.write("\n")
myFile1.write(str(d))
myFile2.write(str(e))
myFile3.write(str(e))
```

The program code for classification is given below. We used R programming language for training and testing ANN classifiers. The training and testing files are given as inputs to the nnet() function and the prediction probabilities returned by the predict() function are averaged together. If the average probability value exceeds 0.5 then, the applications are treated as malware.

```
install.packages('nnet')
library(nnet)
a <- read.csv('eigentrain.csv')
b <- read.csv('eigentest.csv')
a1 <- read.csv('betweennesstrain.csv')
b1 <- read.csv('betweennesstest.csv')
a2 <- read.csv('closenesstrain.csv')
b2 <- read.csv('closenesstest.csv')
c <- nnet(Application~.,size=5,a)
c1 <- nnet(Application~.,size=5,a1)
c2 <- nnet(Application~.,size=5,a2)
p <- predict(c,b)
p1 <- predict(c1,b1)
p2 <- predict(c2,b2)
p3 <- (p+p1+p2)/3
#True Positives
length(which(p3[0:325] > .5))
#False Positives
length(which(p3[326:650] > .5))
```

Source Codes of Chapter 6

The code of GCN is given below. Packages such as Keras and Tensorflow are used to code GCN. We used the implementation of[87] to build our GCN model.

layer.py: The layer.py file is used to create different layers of neural network. The code is given below.

```python
import tensorflow as tf
import keras as k
from utils import *

_LAYER_UIDS = {}
def get_layer_uid(layer_name=''):
    if layer_name not in _LAYER_UIDS:
        _LAYER_UIDS[layer_name] = 1
        return 1
     else:
        _LAYER_UIDS[layer_name] += 1
        return _LAYER_UIDS[layer_name]

def sparse_dropout(x, keep_prob, noise_shape):
    random_tensor = keep_prob
    random_tensor += tf.random_uniform(noise_shape)
    dropout_mask = tf.cast(tf.floor(random_tensor),
    dtype=tf.bool)
    pre_out = tf.sparse_retain(x, dropout_mask)
    return pre_out * (1./keep_prob)

def dot(x, y, sparse=False):
    if sparse:
        res = tf.sparse_tensor_dense_matmul(x, y)
    else:
        res = tf.matmul(x, y)
    return res
class Layer(object):
    def __init__(self, **kwargs):
        layer = self.__class__.__name__.lower()
        name = layer + '_' + str(get_layer_uid(layer))
        self.name = name
        self.weights = {}
        self.sparse_inputs = False
    def _call(self, inputs):
        return inputs
    def __call__(self, inputs):
        with tf.name_scope(self.name):
         outputs = self._call(inputs)
         return outputs

class ConvolutionalLayer(Layer):
    def __init__(self, input_dim, output_dim,
    placeholders, dropout,
```

```
            sparse_inputs, activation, isLast=False,
        bias=False,
        featureless=False, **kwargs):
            super(ConvolutionalLayer, self).
            __init__(**kwargs)
            if dropout:
                self.dropout = placeholders['dropout']
            else:
                self.dropout = 0.
            self.featureless = featureless
            self.activation = activation
            self.support = placeholders['support']
            self.sparse_inputs = sparse_inputs
            self.bias = bias

            # helper variable for sparse dropout
            self.num_features_nonzero =
            placeholders['num_features_nonzero']

            with tf.variable_scope(self.name + '_weights'):
                for i in range(len(self.support)):
                    self.weights['weights_' + str(i)]=
                    glorot([input_dim, output_dim],
                    name='weights_' + str(i))
                if self.bias:
                    self.weights['bias'] =zeros([output_dim],
                    name='bias')

        def _call(self, inputs):
            x = inputs
            # dropout
            if self.sparse_inputs:
                x = sparse_dropout(x, 1-self.dropout,
                self.num_features_nonzero)
            else:
                x = tf.nn.dropout(x, 1-self.dropout)
            # convolve
            supports = list()
            for i in range(len(self.support)):
                if not self.featureless:
                    pre_sup = dot(x, self.weights
                    ['weights_' +
                    str(i)],
                    sparse=self.sparse_inputs)
                else:
```

```python
        pre_sup = self.weights
            ['weights_' + str(i)]
        support = dot(self.support[i],
        pre_sup, sparse=True)
        supports.append(support)
    output = tf.add_n(supports)

    # bias
    if self.bias:
        output += self.weights['bias']
        return self.activation(output)
class PoolingLayer(Layer):
def __init__(self, num_graphs, num_nodes, idx,
input_dim,
output_dim, placeholders, sparse_inputs,activation,
isLast=False, bias=False,featureless=False,
**kwargs):
    super(PoolingLayer, self).__init__(**kwargs)
    self.num_nodes = num_nodes
    self.num_graphs = num_graphs
    self.activation = activation
    self.output_dim = output_dim
    self.input_dim = input_dim
    self.idx = idx

def _call(self, inputs):
    #pooling_matrix = 0
    pooling_matrix = np.array([[0. for i in range
    (self.num_nodes)]
    for k in range(self.num_graphs)])
    idx_aug = np.append(self.idx, self.num_nodes-1)
    idx_aug = idx_aug.astype(int)

    for i in range(self.num_graphs):
        pooling_matrix[i, range(idx_aug[i],
        idx_aug[i+1])] =
        (1/(idx_aug[i+1]-idx_aug[i])
    output = dot(tf.cast(pooling_matrix,
    tf.float32), inputs,
    sparse = False)
    return self.activation(output)
```

fileutils.py: The fileutils.py file provides various utilities to the GCN classifier. This file contains code for loading the inputs and processing them for learning. For this, the inputs of the GCN such as feature matrix, graph label, etc. are converted to numpy arrays (.npz format). The code of fileutils.py is given below.

```python
import numpy as np
import os
from collections import defaultdict
from itertools import groupby
import scipy.sparse as sp
import random
import sys

def parse_index_file(filename):
    index = []
    for line in open(filename):
        index.append(int(line.strip()))
    return index

def encode_onehot(labels):
    classes = set(labels)
    classes_dict = {c: np.identity
    (len(classes))[i, :]
    for i, c in enumerate(classes)}
    onehot = np.array
    (list(map(classes_dict.get, labels)),
    dtype=np.int32)
    return onehot

def sample_mask(idx, l):
    mask = np.zeros(l)
    mask[idx] = 1
    return np.array(mask, dtype=np.bool)

def load_data_outer(dataset, model):
    if model == 'gcn':
        return load_data_gcn(dataset)

def add_one_by_one(l):
    new_l = []
    cumsum = 0
    for elt in l:
        cumsum += elt
        new_l.append(cumsum)
    return new_l

def load_data_gcn(dataset="malware"):
```

```
print('Loading: {} dataset...'.format(dataset))
feats_and_labels = np.genfromtxt("malware\{}.
content".format
(dataset),dtype=np.dtype(str))
features = sp.csr_matrix(feats_and_labels
[:, 1:-1],
dtype=np.int32)
labels = encode_onehot(feats_and_labels
[:, -1])
idx = np.array(feats_and_labels[:, 0],
dtype=np.int32)
idx_map = {j: i for i, j in enumerate(idx)}
edges_unordered = np.genfromtxt("cora\{}.cites".
format(dataset),
dtype=np.int32)
edges = np.array(list(map(idx_map.get,
edges_unordered.flatten())),
dtype=np.int32).reshape(edges_unordered.shape)
adj = sp.coo_matrix((np.ones(edges.shape[0]),
(edges[:, 0], edges[:, 1])),
shape=(labels.shape[0],
labels.shape[0]), dtype=np.int32)
adj = adj + adj.T.multiply(adj.T > adj) -
adj.multiply(adj.T > adj)
print('{} has {} nodes, {} edges,
{} features.'.format(dataset, adj.shape[0],
edges.shape[0],
features.shape[1]))
return features, adj, labels


def get_splits(labels, train_dim, val_dim, test_dim):
    train_ind = range(train_dim[1])
    val_ind = range(val_dim[0], val_dim[1])
    test_ind = range(test_dim[0], test_dim[1])
    labels_train = np.zeros(labels.shape,
    dtype=np.int32)
    labels_val = np.zeros(labels.shape,
    dtype=np.int32)
    labels_test = np.zeros(labels.shape,
    dtype=np.int32)
    labels_train[train_ind] = labels[train_ind]
    labels_val[val_ind] = labels[val_ind]
    labels_test[test_ind] = labels[test_ind]
    train_mask = sample_mask(train_ind,
    labels.shape[0])
```

```
    val_mask = sample_mask(val_ind,
    labels.shape[0])
    test_mask = sample_mask(test_ind,
    labels.shape[0])
    return labels_train, labels_val,
    labels_test,
    train_ind, val_ind,
    test_ind, train_mask,
    val_mask, test_mask

def get_splits_graphs_basic(num_graphs, labels,
train_dim,
val_dim,
test_dim, oldidx):
    idx = np.array([i for i in range(num_graphs)])
    train_ind = [idx[train_dim[0] : train_dim[1]]]
    val_ind = [idx[val_dim[0] : val_dim[1]]]
    test_ind = [idx[test_dim[0] : test_dim[1]]]
    labels_train = np.zeros(labels.shape, dtype=np.int32)
    labels_val = np.zeros(labels.shape, dtype=np.int32)
    labels_test = np.zeros(labels.shape, dtype=np.int32)
    labels_train[train_ind] = labels[train_ind]
    labels_val[val_ind] = labels[val_ind]
    labels_test[test_ind] = labels[test_ind]
    train_mask = sample_mask(train_ind, labels.shape[0])
    val_mask = sample_mask(val_ind, labels.shape[0])
    test_mask = sample_mask(test_ind, labels.shape[0])
    return labels_train, labels_val, labels_test,
    train_ind,
    val_ind, test_ind, train_mask, val_mask, test_mask

def get_splits_graphs(num_graphs, labels, train_dim,
val_dim, test_dim, idx):
    idx_incr = np.array([i for i in range(num_graphs)])
    idx_incr = idx + idx_incr
    random.shuffle(idx_incr)
    train_ind = [idx_incr[train_dim[0] : train_dim[1]]]
    val_ind = [idx_incr[val_dim[0] : val_dim[1]]]
    test_ind = [idx_incr[test_dim[0] : test_dim[1]]]
    labels_train = np.zeros(labels.shape, dtype=np.int32)
    labels_val = np.zeros(labels.shape, dtype=np.int32)
    labels_test = np.zeros(labels.shape, dtype=np.int32)
    labels_train[train_ind] = labels[train_ind]
    labels_val[val_ind] = labels[val_ind]
    labels_test[test_ind] = labels[test_ind]
```

```
    train_mask = sample_mask(train_ind, labels.shape[0])
    val_mask = sample_mask(val_ind, labels.shape[0])
    test_mask = sample_mask(test_ind, labels.shape[0])
    return labels_train, labels_val, labels_test,
    train_ind, val_ind,
    test_ind, train_mask, val_mask, test_mask

def load_data_basic(num_nodes, num_graphs,
num_classes,dim_feats,dataset_name):
    global_nodes_idx = find_insert_position(dataset_name)
    adj_matrix = build_adj_diag_basic(num_nodes, num_graphs,
    dataset_name)
    node_feats = build_feats_basic(num_nodes, num_graphs,
    dim_feats, dataset_name)
    graph_labels = build_labels_basic(num_graphs, num_classes,
    num_nodes,dataset_name)
    return adj_matrix, node_feats, graph_labels,
    global_nodes_idx

def load_data(num_nodes, num_graphs, num_classes, dim_feats,
dataset_name):
    global_nodes_idx = find_insert_position(dataset_name)
    adj_matrix = build_adj_diag(num_nodes, num_graphs,
    global_nodes_idx,dataset_name)
    node_feats = build_feats_vertConc(global_nodes_idx,
    num_nodes,num_graphs, dim_feats,
    dataset_name)
    graph_labels = build_labels_vertConc(num_graphs,
    num_classes,num_nodes,global_nodes_idx,
    dataset_name)
    return adj_matrix, node_feats, graph_labels,
    global_nodes_idx


def find_insert_position(dataset_name):
    path = dataset_name +"/"+dataset_name.upper()
    +"_graph_indicator.txt"
    node_ind = np.genfromtxt(path)
    fake_idx = np.where(node_ind[:-1] != node_ind[1:])[0]
    fake_idx = fake_idx + 1
    fake_idx = np.insert(fake_idx, 0, 0)
    return fake_idx

def build_labels_basic(graphs, num_classes, num_nodes,
dataset_name ):
```

```
    if(os.path.exists(dataset_name +
    "_labels_basic.npy")):
    labels = np.load(dataset_name +"_labels_basic.npy")
    return labels
    path=dataset_name +"/"+dataset_name.upper()+
    "_graph_labels.txt"
    true_labels = np.loadtxt(path
    , dtype='i', delimiter=',')
    labels = encode_onehot(true_labels)
    np.save(dataset_name +"_labels_basic.npy", labels)
    return labels

def build_labels_vertConc(graphs, num_classes, num_nodes,
idx, dataset_name ):
    if(os.path.exists(dataset_name +"_labels_aug.npy")):
        labels = np.load(dataset_name +"_labels_aug.npy")
        return labels
    path=dataset_name +"/"+dataset_name.upper()+
    "_graph_labels.txt"
    true_labels = np.loadtxt(path, dtype='i',
    delimiter=',')
    true_labels = encode_onehot(true_labels)
    labels = np.array([[0 for i in range(num_classes)]
    for k in range(num_nodes)])
    for i in range(graphs):
        labels = np.insert(labels, idx[i]+i,
        true_labels[i], axis=0)
        print("row: ")
        print(i)
    np.save(dataset_name +"_labels_aug.npy", labels)
    return labels

def build_feats_basic(num_nodes, graphs, dim_feats,
dataset_name):
    if(os.path.exists(dataset_name+
    "_feats_matrix_basic.npz")):
        feats_matrix=sp.load_npz(dataset_name+
        "_feats_matrix_basic.npz")
        return feats_matrix
    path=dataset_name +"/"+dataset_name.upper()+
    "_node_attributes.txt"
    feats_matrix = np.loadtxt(path, delimiter=',')
    feats_matrix = sp.csr_matrix(feats_matrix)
    sp.save_npz(dataset_name+"_feats_matrix_basic",
    feats_matrix)
```

```
    return feats_matrix

def build_feats_vertConc(idx, num_nodes, graphs, dim_feats,
dataset_name):
    if(os.path.exists(dataset_name
    +"_feats_matrix_aug.npz")):
        feats_matrix = sp.load_npz(dataset_name+

        "_feats_matrix_aug.npz")
        return feats_matrix
    path=dataset_name +"/"+dataset_name.upper()+
    "_node_attributes.txt"
    feats_matrix = np.loadtxt(path, delimiter=',')
    fake_feats = np.array([[0. for i in range(dim_feats)]
    for k in range(graphs)])
    print("inserting global node rows:")
    for i in range(graphs):
        feats_matrix = np.insert(feats_matrix, idx[i]+i,
        fake_feats[i],
        axis=0)
        print("row:")
        print(i)
    feats_matrix = np.absolute(feats_matrix)
    feats_matrix = sp.csr_matrix(feats_matrix)
    sp.save_npz(dataset_name+"_feats_matrix_aug",
    feats_matrix)
    return feats_matrix

def build_feats_vertConc_mean_features(idx,num_nodes,
graphs,dim_feats, dataset_name):
    if(os.path.exists(dataset_name+
    "_feats_matrix_aug_with_mean.npz")):
        feats_matrix = sp.load_npz(dataset_name+
        "_feats_matrix_aug_with_mean.npz")
        return feats_matrix
    path=dataset_name +"/"+dataset_name.upper()+
    "_node_attributes.txt"
    aug_idx = np.append(idx, int(num_nodes-1))
    feats_matrix = np.loadtxt(path, delimiter=',')
    fake_feats = np.array([[ np.mean(feats_matrix
    [[ range(aug_idx[k],aug_idx[k+1]) ] ,[i] ] )
    for i in range(dim_feats)] for k in range(graphs)])
    print("inserting global node rows:")
    for i in range(graphs):
        feats_matrix = np.insert(feats_matrix, idx[i]+i,
```

```
                    fake_feats[i],
                    axis=0)
                print("row: ")
                print(i)
            feats_matrix = sp.csr_matrix(feats_matrix)
            sp.save_npz(dataset_name+"_feats_matrix_aug_with_mean",
            feats_matrix)
            return feats_matrix


def build_adj_diag_basic(nodes, graphs, dataset_name):
    if(os.path.exists(dataset_name +"_adj_matrix_basic.npz")):
        adj_matrix = sp.load_npz(dataset_name+
        "_adj_matrix_basic.npz")
        return adj_matrix

    path=dataset_name +"/"+dataset_name.upper()+"_A.txt"
    tmpdata = np.genfromtxt(path, dtype=np.dtype(str))
    ind1 = tmpdata[:, 1]
    ind2 = tmpdata[:, 0]
    adj_matrix = [[0 for i in range(nodes)]
    for k in range(nodes)]
    for i in range(len(ind1)):
        print(i)
        u = ind1[i]
        v = ind2[i]
        u = int(u)
        v = int(v[:-1])
        adj_matrix[u-1][v-1] = 1

    adj_matrix = np.matrix(adj_matrix)
    adj_matrix = sp.coo_matrix(adj_matrix)
    sp.save_npz(dataset_name + "_adj_matrix_basic", adj_matrix)
    return adj_matrix


def build_adj_diag(nodes, graphs, idx, dataset_name):
    if(os.path.exists(dataset_name +"_adj_matrix_aug.npz")):
        adj_matrix = sp.load_npz(dataset_name+
        "_adj_matrix_aug.npz")
        return adj_matrix
    path=dataset_name +"/"+dataset_name.upper()+"_A.txt"
    nodes_tot = nodes+graphs
    fake_matrix = np.array([[0 for i in range(nodes)]
    for k in range
    (graphs)])
```

```
node_ind = np.genfromtxt(dataset_name+"/"
+dataset_name.upper()+
"_graph_indicator.txt")
node_ind = node_ind.tolist()
occ = [len(list(group)) for key, group in groupby(node_ind)]
occ = add_one_by_one(occ)
occ.insert(0, 1)
ranges = list(zip(occ[1:], occ))
upper_idx, lower_idx = map(list, zip(*ranges))

for index in range(graphs):
    fake_matrix[index][(lower_idx[index] -1) :
    (upper_idx[index]-1)] = 1
print("parsing original adj matrix")
tmpdata = np.genfromtxt(path, dtype=np.dtype(str))
ind1 = tmpdata[:, 1]
ind2 = tmpdata[:, 0]
adj_matrix = [[0 for i in range(nodes)]
for k in range(nodes)]
for i in range(len(ind1)):
    u = ind1[i]
    v = ind2[i]
    u = int(u)
    v = int(v[:-1])
    adj_matrix[u-1][v-1] = 1
print("inserting global node rows:")
for i in range(graphs):
    adj_matrix = np.insert(adj_matrix, idx[i]+i,
    fake_matrix[i], axis=0)
    print("row: ")
    print(i)
lower_idx_new = [0 for i in range(graphs)]
upper_idx_new = [0 for i in range(graphs)]
for i in range(graphs):
    lower_idx_new[i] = lower_idx[i]+i
    upper_idx_new[i] = upper_idx[i]+i
vert_padding = np.array([[0 for i in range(nodes_tot)]
for k
in range(graphs)])
for i in range(graphs):
    vert_padding[i][(lower_idx_new[i]-1):
    (upper_idx_new[i]-1)] = 1
print("inserting global node columns:")
for i in range(graphs):
    adj_matrix = np.insert(adj_matrix, idx[i]+i,
```

```
        vert_padding[i], axis=1)
        print("column: ")
        print(i)
    adj_matrix = np.matrix(adj_matrix)
    adj_matrix = sp.coo_matrix(adj_matrix)
    sp.save_npz(dataset_name + "_adj_matrix_aug", adj_matrix)
    return adj_matrix
```

neural networks.py: The neural network.py contains code of how the GCN learns the system call graphs to classify whether the system call graph is malicious or not.

```
from layers import *
from utils import *
import tensorflow as tf
flags = tf.app.flags
FLAGS = flags.FLAGS
class BaseNet(object):
    def __init__(self, **kwargs):
        self.name = self.__class__.__name__.lower()
        self.weights = {}
        self.placeholders = {}
        self.layers = []
        self.activations = []
        self.inputs = None
        self.outputs = None
        self.loss = 0
        self.accuracy = 0
        self.optimizer = None
        self.opt_op = None

    def _build(self):
        raise NotImplementedError
    def build(self):
        with tf.variable_scope(self.name):
            self._build()
        self.activations.append(self.inputs)
        for layer in self.layers:
            hidden = layer(self.activations[-1])
            self.activations.append(hidden)
        self.outputs = self.activations[-1]
        self.weights = {var.name: var for var in
        tf.get_collection(tf.GraphKeys.GLOBAL_VARIABLES,
        scope=self.name)}
        self._loss()
```

```
        self._accuracy()
        self.opt_op = self.optimizer.minimize(self.loss)
    def predict(self):
        pass
    def _loss(self):
        raise NotImplementedError
    def _accuracy(self):
        raise NotImplementedError

class GCN(BaseNet):
    def __init__(self, placeholders, input_dim, **kwargs):
        super(GCN, self).__init__(**kwargs)
        self.inputs = placeholders['feats']
        self.input_dim = input_dim
        self.output_dim = placeholders['labels'].
        get_shape().as_list()[1]
        self.placeholders = placeholders
        self.optimizer = tf.train.AdamOptimizer
        (learning_rate=FLAGS.learning_rate)
        self.build()

    def _loss(self):
        # Weight decay loss
        for var in self.layers[0].weights.values():
            self.loss += FLAGS.weight_decay *
            tf.nn.l2_loss(var)
        #cross entropy loss
        self.loss += masked_cross_entropy
        (self.outputs, self.placeholders
        ['labels'],

        self.placeholders['labels_mask'])

    def _accuracy(self):
        self.accuracy = masked_accuracy(self.outputs,

        self.placeholders['labels'],

        self.placeholders['labels_mask'])

    def _build(self):
        self.layers.append(ConvolutionalLayer
        (input_dim=self.input_dim,output_dim=FLAGS.hidden1,
        placeholders=self.placeholders,
        activation=tf.nn.relu,dropout=True,
```

```python
                sparse_inputs=True,featureless=False))
            self.layers.append(ConvolutionalLayer
            (input_dim=FLAGS.hidden1,
            output_dim=self.output_dim,
            placeholders=self.placeholders,
            activation=lambda x: x,
            dropout=True,
            sparse_inputs=False))


    def predict(self):
        return tf.nn.softmax(self.outputs)


class GCNGraphs(BaseNet):
    def __init__(self, placeholders, input_dim, featureless,
    idx, num_graphs, num_nodes,
    with_pooling, **kwargs):
        super(GCNGraphs, self).__init__(**kwargs)
        self.pooling = with_pooling
        self.num_graphs = num_graphs
        self.num_nodes = num_nodes
        self.idx = idx
        self.inputs = placeholders['feats']
        self.input_dim = input_dim
        self.output_dim = placeholders['labels'].get_shape().
        as_list()[1]
        self.placeholders = placeholders
        self.featureless = featureless
        self.optimizer = tf.train.AdamOptimizer
        (learning_rate=FLAGS.learning_rate)
        self.build()

    def _loss(self):
        # Weight decay loss
        for var in self.layers[0].weights.values():
            self.loss += FLAGS.weight_decay *
            tf.nn.l2_loss(var)

        #cross entropy loss
        self.loss += masked_cross_entropy
        (self.outputs,
        self.placeholders['labels'],
        self.placeholders['labels_mask'])
```

```
def _accuracy(self):
    self.accuracy = masked_accuracy(self.outputs,
    self.placeholders['labels'],
    self.placeholders['labels_mask'])

def _build(self):
    self.layers.append(ConvolutionalLayer(input_dim=
    self.input_dim,
    output_dim=FLAGS.hidden2,
    placeholders=self.placeholders,
    activation=tf.nn.relu,
    dropout=True,
    sparse_inputs=True,
    featureless = self.featureless))
    self.layers.append(ConvolutionalLayer(input_dim=
    FLAGS.hidden2,
    output_dim=self.output_dim,
    placeholders=self.placeholders,
    activation=lambda x: x,
    dropout=True,
    sparse_inputs=False,
    featureless = False))
    if self.pooling:
        self.layers.append(PoolingLayer(num_graphs =
        self.num_graphs,
        num_nodes = self.num_nodes,
        idx=self.idx,
        input_dim=self.output_dim,
        output_dim=self.output_dim,
        placeholders=self.placeholders,
        activation=lambda x: x,
        sparse_inputs=False,
        featureless = False))
def predict(self):
    return tf.nn.softmax(self.outputs)
```

Graph Classification Using GCN.py: The graph classification using GCN.py file contains the code for specifying the parameters used for learning the graph representation such as the number of epochs, learning rate, etc. The code of graph classification using GCN.py is given below.

```
from __future__ import division
from __future__ import print_function
import time
import tensorflow as tf
```

```python
import matplotlib.pyplot as plt
from sklearn.metrics import roc_curve
from sklearn.metrics import roc_auc_score
from file_utils import *
from utils import *
from neural_networks import GCNGraphs
from neural_networks import GCN
# Set random seed
seed = 123
np.random.seed(seed)
tf.set_random_seed(seed)
def del_all_flags(FLAGS):
    flags_dict = FLAGS._flags()
    keys_list = [keys for keys in flags_dict]
    for keys in keys_list:
        FLAGS.__delattr__(keys)


del_all_flags(tf.flags.FLAGS)


# Settings
flags = tf.app.flags
FLAGS = flags.FLAGS
flags.DEFINE_string('dataset', 'MALWARE',
'which dataset to
load')
flags.DEFINE_boolean('with_pooling', True,
'if a mean value for
graph labels is computed via pooling(True)
or via global nodes
(False)')
flags.DEFINE_boolean('featureless', False,
'If nodes are featureless')
#only if with_pooling = False
flags.DEFINE_float('learning_rate', 0.01,
'Initial learning rate.')
flags.DEFINE_integer('epochs', 200,
'Number of epochs to train.')
flags.DEFINE_integer('hidden1', 32, 'Number of units in
hidden layer 1.')
flags.DEFINE_integer('hidden2', 64, 'Number of units in
hidden layer 2.')
flags.DEFINE_integer('hidden3', 16, 'Number of units in
hidden layer 3.')
flags.DEFINE_float('dropout', 0, 'Dropout rate
```

```
(1 - keep probability).')
flags.DEFINE_float('weight_decay', 5e-4,
'Weight for L2 loss on embedding
matrix.')
flags.DEFINE_integer('early_stopping', 10,
'Tolerance for early stopping (# of epochs).')

elif FLAGS.dataset=='MALWARE':
    num_nodes = 55406
    num_graphs = 2131
    tot = 57537
    num_classes = 2
    num_feats = 5
    dataset_name = "malware"
    splits = [[0,1400], [1400, 1500], [2000, 2131]]

if not FLAGS.with_pooling:
    adj, features, labels, idx = load_data(num_nodes,
    num_graphs,
    num_classes, num_feats, dataset_name)
    y_train, y_val, y_test, idx_train, idx_val, idx_test,
    train_mask,
    val_mask, test_mask = get_splits_graphs(num_graphs,
    labels,
    splits[0], splits[1], splits[2], idx)

else:
    adj, features, labels, idx = load_data_basic(num_nodes,
    num_graphs, num_classes, num_feats, dataset_name)
    y_train, y_val, y_test, idx_train,
    idx_val, idx_test, train_mask, val_mask, test_mask=
    get_splits_graphs_basic(num_graphs,
    labels, splits[0], splits[1], splits[2], idx)
support = [preprocess_adj(adj, True, False)]
features = process_features(features)

num_supports = 1

GCN_placeholders = {
    'idx' :tf.placeholder(tf.int32),
    'support': [tf.sparse_placeholder(tf.float32)
    for i in range(num_supports)],
    'feats': tf.sparse_placeholder(tf.float32,
    shape=tf.constant(features[2],
    dtype=tf.int64)),
```

```
    'labels': tf.placeholder(tf.float32, shape=(None,
    y_train.shape[1])),
    'labels_mask': tf.placeholder(tf.int32),
    'dropout': tf.placeholder_with_default(0., shape=()),
    'num_features_nonzero': tf.placeholder(tf.int32),
    # helper variable for sparse dropout
}

if FLAGS.featureless:
    if not FLAGS.with_pooling:
        input_dim = tot
    else:
        input_dim = num_nodes
else:
    input_dim = features[2][1]

# Create network
featureless = (FLAGS.featureless)
network = GCNGraphs(GCN_placeholders, input_dim,
featureless, idx, num_graphs, num_nodes,
FLAGS.with_pooling)

# Initialize session
sess = tf.Session()
# Init variables
sess.run(tf.global_variables_initializer())
cost_val = []
train_dict = build_dictionary_GCN(features, support,
y_train,
train_mask, GCN_placeholders)
train_dict.update({GCN_placeholders['dropout']:
FLAGS.dropout})
train_loss = [0. for i in range(0, FLAGS.epochs)]
val_loss = [0. for i in range(0, FLAGS.epochs)]

def evaluate(features, support, labels, mask,
placeholders):
    t_test = time.time()
    feed_dict_val = build_dictionary_GCN
    (features, support,
    labels, mask, GCN_placeholders)
    outs_val = sess.run([network.loss, network.accuracy],
    feed_dict=feed_dict_val)
    return outs_val[0], outs_val[1], (time.time()
    - t_test)
```

```
# Train network
for epoch in range(FLAGS.epochs):
    t = time.time()
    # Training step
    train_out = sess.run([network.opt_op,
    network.loss, network.accuracy,
    network.outputs],
    feed_dict=train_dict)
    train_loss[epoch] = train_out[1]

    # Validation
    t_test = time.time()

    cost, acc, duration = evaluate(features,
    support, y_val, val_mask, GCN_placeholders)
    cost_val.append(cost)
    val_loss[epoch] = cost
    print("Epoch:", '%04d' % (epoch + 1),
    "train_loss=",
    "{:.5f}".format(train_out[1]),
     "train_acc=", "{:.5f}".format(train_out[2]),
     "val_loss=","{:.5f}".format(cost),
     "val_acc=", "{:.5f}".format(acc),
     "time=", "{:.5f}".format(time.time() - t))

    """ if epoch > FLAGS.early_stopping and
    cost_val[-1] >
    np.mean(cost_val[-(FLAGS.early_stopping+1):-1]):
        print("Early stopping...")
        break """
#network.save(sess,'mymodel')
print("Optimization Finished!")

test_cost, test_acc, test_duration =
evaluate(features,
support, y_test, test_mask, GCN_placeholders)
print("Test set results:", "cost=", "{:.5f}".
format(test_cost),
      "accuracy=",
      "{:.5f}".format
      (test_acc),
      "time=", "{:.5f}".format(test_duration))
epochs = [i for i in range(0, FLAGS.epochs)]
plt.plot(np.array(epochs),
```

```
np.array(train_loss), color='g')
plt.plot(np.array(epochs),
np.array(val_loss), color='orange')
plt.xlabel('Epochs')
plt.ylabel('Loss')
plt.title('Train and validation(yellow)
loss over epochs with {}
dataset'.format(dataset_name))
plt.show()
probs = model.predict_proba(x_test)
probs = probs[:, 1]
auc = roc_auc_score(y_test, probs)
print('AUC: %.3f' % auc)
fpr, tpr, thresholds = roc_curve(y_test, probs)
plt.plot([0, 1], [0, 1], linestyle='--')
# plot the roc curve for the model
plt.plot(fpr, tpr, marker='.')
# show the plot
plt.show()
```

Source Codes of Chapter 7

The python code for graph signal construction is given below. This graph signal construction code receives system call sequences of several applications as input and gives the graph signal vectors as the output. In this process, the python code first preprocess the system call sequences by eliminating arguments and irrelevant system calls. Then, it assigns alternative name to system calls instead of their original name for convenience. After that, it extracts system call count values and adjacency matrix from each system call sequence. It then computes the transformed graph signals and save it in the csv file for machine learning classification.

```
import csv
import networkx as nx
import math
from sklearn import preprocessing
import numpy
import os, sys
import re
import numpy as np
path = "C:\\Dataset"
myFile1 = open('C:\\GraphSignals\\graphsignals.csv',
'w')
```

```
files=os.listdir(path)
z=[]
z57=[]
for Files in files:
    # Reading system call sequence
    with open(path+"/"+Files) as f:
        for line in f:
            z=line.split("(")[0]
            z57=z57+z
    signals=[z57.count('A'),z57.count('B'),
    z57.count('C'),
    z57.count('D'),z57.count('E'),z57.count('F'),
    z57.count('G'),z57.count('H'),z57.count('I'),
    z57.count('J'),z57.count('K'),z57.count('L'),
    z57.count('M'),z57.count('N'),z57.count('O'),
    z57.count('P'),z57.count('Q'),z57.count('R'),
    z57.count('S'),z57.count('T'),z57.count('U'),
    z57.count('V'),z57.count('W'),z57.count('X'),
    z57.count('Y'),z57.count('Z')]
    for i in range(0,26):
        if sum(signals)!=0:
            signals.append(signals[i]/sum(signals))
        else:
            signals.append(signals[i]/1)
    #Create adjacency matrix
    def rank(c):
        return ord(c) - ord('A')
    T = [rank(c) for c in z9]
    M = [[0]*26 for _ in range(26)]
    for (i,j) in zip(T,T[1:]):
        M[i][j] += 1
    for row in M:
        n = sum(row)
        if n > 0:
            row[:] = [f/sum(row) for f in row]
    #Transformed Graph Signals construction from
    adjacency matrix
        transformed_signals=np.zeros(26)
    for i in range(0,26):
        for j in range(0,26):
        transformed_signals[i]=p[i]+(M[i][j]*z12345[j])
# Logging graph signals as csv files
    myFile1.write("\n")
    myFile1.write(str(transformed_signals))
```

The python codes for the classifiers are given below. The file graphsignals.csv is split in the ratio 9:1 for training and testing. That is 90% of samples are used for training and the remaining 10% samples are used for testing. These training and test data sets are given as inputs to various ML algorithms for determining their performance. Here, label_pred is the output (predicted class) of the ML classifier.

```python
from sklearn import svm
import pandas as pd
from sklearn.naive_bayes import GaussianNB
from sklearn.ensemble import RandomForestRegressor
from sklearn.tree import DecisionTreeClassifier
from sklearn.neural_network import MLPClassifier
data=pd.read_csv('graphsignals.csv')
graphsignals.target=data.labels
graphsignals.data=data.drop('target',axis=1)
# 90% training and 30% test
Data_train, Data_test, label_train, label_test =
train_test_split(graphsignals.data,
graphsignals.target, test_size=0.1,random_state=109)
#SVM Classifier
svmcl = svm.SVC(kernel='linear')
svmcl.fit(Data_train, label_train)
label_pred = clf.predict(Data_test)
#Naive bayes Classifier
NBC = GaussianNB()
NBC.fit(Data_train, label_train)
label_pred  =  classifier.predict(Data_test)
#Decision Tree Classifier
DT=DecisionTreeClassifier()
DT.fit(Data_train,label_train)
label_pred = DT.predict(Data_test)
#Random Forest Classifier
RFC = RandomForestRegressor(n_estimators = 1000,
random_state = 42)
RFC.fit(Data_train, label_train)
label_pred=rf.predict(Data_test)
#ANN Classifier
clf = MLPClassifier(solver='lbfgs',
alpha=1e-5,hidden_layer_sizes=(5, 2), random_state=1)
clf.fit(Data_train, label_train)
label_pred=clf.predict(Data_test)
```

Source Codes of Chapter 8

The python code for extracting the system call pattern of an application is given below.

```python
import numpy
import os, sys
import re
from numpy.linalg import matrix_power
from numpy.linalg import matrix_rank
import numpy as np
path = "C:\\Dataset"
files=os.listdir(path)
z=[]
z57=[]
for Files in files:
    # Reading system call sequence
    with open(path+"/"+Files) as f:
        for line in f:
            a=line.split("(")[0]
            z57=z57+a

    #Create adjacency matrix
    def rank(c):
        return ord(c) - ord('A')
    T = [rank(c) for c in z57]
    M = [[0]*26 for _ in range(26)]
    for (i,j) in zip(T,T[1:]):
        M[i][j] += 1
    for row in M:
        n = sum(row)
        if n > 0:
            row[:] = [f/sum(row) for f in row]
    res = re.findall(r'\z.*?\z', z57)
    j=1
    mat=M
    while mat_rank(mat)==1:
        mat=matrix_power(M,j)
        j=j+1
    dist=Mat[1,]
    # Entropy Calculation
    entr=0
    for i in range(0,26):
        for j in range(0,26):
            entr=entr+dist[i]*M[i,j]*math.log(M[i,j],
```

```
                10)
    # Subsequence Probability Calculation
    for i in range(0,len(res)):
        distance[i]=1
        for j in range(0,len(res[i])):
            distance[i]=distance[i]*z57.count
            (res[i][j])
            /len(z57)
        distance[i]=-1/26*math.log(distance[i])
    # LTP Extraction
    minvalue=1
    for i in range(0,len(res)):
        minp[i]=abs(distance[i]-entr)
        if minp[i]<minvalue:
            LTP=minp[i]
```

The python code for the Jaro-winkler similarity matching for the system call patterns is given below.

```
import numpy
import os, sys
import re
import numpy as np
from pyjarowinkler import distance
path = "C:\\Dataset"
files=os.listdir(path)
z=[]
z57=[]
i=1
malware=false
# Reading the LTPs
with open("C:\\pattern.txt") as p:
    for line in p:
        pattern[i]=line
        i=i+1
for Files in files:
    # Reading system call sequence
    with open(path+"/"+Files) as f:
        for line in f:
            z=line.split("(")[0]
            z57=z57+z
        res = re.findall(r'\z.*?\z', z57)
        # Pattern matching using Jaro-Winkler similarity
        metric
        for i in range(0,len(pattern)):
            for j in range(0,len(res)):
```

```
            if distance.get_jaro_distance
            (pattern[i], res[j],
            winkler=True, scaling=0.1)>0.85 :
                print ("Malware Application")
                malware=true
                break
    if malware:
        break
```

Bibliography

[1] 'FakeInstaller' Leads the Attack on Android Phones. `https://www.mcafee.com/blogs/mobile-security/fakeinstaller-leads-the-attack-on-android-phones/`. Accessed on 03-04-2022.

[2] 5 types of Android malware that made headlines in 2017. `https://www.wxii12.com/article/5-types-of-android-malware-that-made-headlines-in-2017/14508001`. Accessed on 03-06-2021.

[3] A Short History of Mobile Malware ProAndroidDev. `https://proandroiddev.com/a-short-history-of-mobile-malware-296570ed5c1b`. Accessed on 03-06-2021.

[4] Analysis of trojan-sms.androidos.fakeplayer. AT & T alien labs. `https://cybersecurity.att.com/blogs/labs-research/analysis-of-trojan-sms.androidos.fakeplayer`. Accessed on 03-06-2021.

[5] Anatomy of Android application. `https://www.tutorialspoint.com/android/android_hello_world_example.html`. Accessed on 03-06-2021.

[6] Android aapt tool. `https://androidaapt.com/`. Accessed on 03-06-2021.

[7] Android dx tool. `https://github.com/rover12421/AndroidDx`. Accessed on 03-06-2021.

[8] Android malware 2019. `https://github.com/sk3ptre/AndroidMalware2019`. Accessed on 03-06-2021.

[9] Android malware samples. `https://github.com/ashishb/android-malware`. Accessed on 11-02-2022.

[10] Android malware takes advantage of covid-19. `https://www.buguroo.com/en/labs/Androidmalwaretakesadvantageofcovid19`. Accessed on 03-06-2021.

[11] Android suffers "Gazon" malware outbreak. `https://www.silicon.co.uk/mobility/mobile-apps/android-mobile-malware`. Accessed on 03-06-2021.

[12] Angry Android hacker hides xbot malware in popular application icons. https://blog.avast.com/2015/02/17/angry-android-hacker-hides-xbot-malware-in-popular-application-icons/. Accessed on 03-06-2021.

[13] API level. http://www.dre.vanderbilt.edu/~schmidt/android/android-4.0/out/target/common/docs/doc-comment-check/guide/appendix/api-levels.html/. Accessed on 27-11-2021.

[14] Application signing. https://source.android.com/security/apksigning. Accessed on 03-06-2021.

[15] Bluetooth-worm: Symbos/cabir description, F-Secure labs. https://www.f-secure.com/v-descs/cabir.shtml. Accessed on 03-06-2021.

[16] Comebot. https://github.com/ashishb/android-malware. Accessed on 03-06-2021.

[17] CovidLock: Android Ransomware Spreading Amid COVID-19 Epidemic. https://cyware.com/research-and-analysis/covidlock-android-ransomware-spreading-amid-covid-19-epidemic-4a5b. Accessed on 03-06-2021.

[18] Dendroid Malware Can Take Over Your Camera, Record Audio, And Sneak Into Google Play. https://www.lookout.com/blog/dendroid. Accessed on 03-06-2021.

[19] Discord, Your Place to Talk and Hang Out. https://discord.com/. Accessed on 03-06-2021.

[20] DroidKungFu and the exploits Rage Against The Cage and Exploid for Android. https://www.malware.unam.mx/en/content/droidkungfu-and-exploits-rageagainstthecage-and-exploid-android. Accessed on 03-06-2021.

[21] Encryption. https://source.android.com/security/encryption. Accessed on 03-06-2021.

[22] Exodus. https://www.sophos.com/en-us/threat-center/threat-analyses/viruses-and-spyware/Exodus. Accessed on 03-06-2021.

[23] Faceapp malware. https://www.forbes.com/sites/kateoflahertyuk/2019/12/03/fbi-faceapp-investigation-confirms-threat-from-apps-developed-in-russia/?sh=45354a5645bc. Accessed on 11-04-2022.

[24] Fake Angry Birds Game spreading Malware from Android Market. https://thehackernews.com/2012/01/fake-angry-birds-game-spreading-malware.html. Accessed on 03-06-2021.

[25] Full report on Cerberus, an Android banking trojan. `https://www.buguroo.com/en/labs/full-report-on-cerberus-an-android-banking-trojan`. Accessed on 03-06-2021.

[26] The fundamental Android security models. `https://medium.com/modulotech/the-fundamental-android-security-models-64f07dda006a`. Accessed on 03-06-2021.

[27] Gingermaster malware seen using root exploit for Android gingerbread threat-post. `https://threatpost.com/gingermaster-malware-seen-using-root-exploit-android-gingerbread-081811/75559/`. Accessed on 03-06-2021.

[28] Google Play Store. `https://play.google.com/store`. Accessed on 21-8-2017.

[29] Hacktivist Android Trojan Designed to Fight App Piracy | PCWorld. `https://www.pcworld.com/article/223842/hacktivist_android_trojan_designed_to_fight_app_piracy.html`. Accessed on 03-06-2021.

[30] How can Android app permissions be exploited by attackers? `https://searchsecurity.techtarget.com/answer/How-can-Android-app-permissions-be-exploited-by-attackers`. Accessed on 03-06-2021.

[31] Intent redirection vulnerabilities in popular Android apps spotlight danger of dynamic code loading, warn researchers. `https://portswigger.net/daily-swig/intent-redirection-vulnerabilities-in-popular-android-apps-spotlight-danger-of-dynamic-code-loading-warn-researchers`. Accessed on 03-06-2021.

[32] Intents and intent filters. `https://developer.android.com/guide/components/intents-filters`. Accessed on 03-06-2021.

[33] Internet of things. `https://en.wikipedia.org/wiki/Internet_of_things`. Accessed on 03-04-2022.

[34] The IoT attack surface: Threats and security solutions. `https://en.wikipedia.org/wiki/Internet_of_things`. Accessed on 03-06-2021.

[35] Joker malware hits Google Play with 17 variants. `https://securityintelligence.com/news/joker-malware-hits-Google-play-with-17-variants/`. Accessed on 11-02-2021.

[36] The layers of the Android security model. `https://proandroiddev.com/the-layers-of-the-android-security-model-90f471015ae6`. Accessed on 03-06-2021.

[37] Malbus: Popular south korean bus app series in google play found dropping malware after 5 years of development | mcafee blogs. `https://www.mcafee.com/blogs/other-blogs/mcafee-labs/malbus-popular-south-korean-bus-app-series-in-Google-play-found-dropping-malware-after-5-years-of-development/`. Accessed on 03-06-2021.

[38] Malware definition. `http://techterms.com/definition/malware`. Accessed on 03-03-2022.

[39] Malware evolution: PC-based vs. mobile. `https://blog.checkpoint.com/2013/01/02/malware-evolution-pc-based-vs-mobile-2/`. Accessed on 03-06-2021.

[40] Mobile Campaign 'Bouncing Golf' Affects Middle East. `https://www.trendmicro.com/en_us/research/19/f/mobile-cyberespionage-campaign-bouncing-golf-affects-middle-east.html`. Accessed on 03-04-2022.

[41] New Android spyware targets users in Pakistan – Sophos News. `https://news.sophos.com/en-us/2021/01/12/new-android-spyware-targets-users-in-pakistan/`. Accessed on 03-06-2021.

[42] New IoT-malware grew three fold in 2018. `https://www.kaspersky.com/about/press-releases/2018_new-iot-malware-grew-three-fold-in-h1-2018`. Accessed on 03-06-2021.

[43] NIST Cybersecurity Framework (CSF). `https://www.gsa.gov/technology/technology-products-services/it-security/nist-cybersecurity-framework-csf`. Accessed on 03-06-2021.

[44] November 2020's most wanted malware: Notorious phorpiex botnet returns as most impactful infection. `https://www.globenewswire.com/news-release/2020/12/09/2142018/0/en/November-2020-s-Most-Wanted-Malware-Notorious-Phorpiex-Botnet-Returns-As-Most-Impactful-Infection.html`. Accessed on 06-04-2022.

[45] Over 14% Indians affected by 'Shopper' malware. `https://www.thehindu.com/sci-tech/technology/internet/over-14-indians-affected-by-shopper-malware/article30555480.ece`. Accessed on 03-06-2021.

[46] Pegasus Spyware and Citizen Surveillance: What You Need to Know. `https://www.cnet.com/tech/mobile/pegasus-spyware-and-citizen-surveillance-what-you-need-to-know/`. Accessed on 03-06-2021.

[47] Ransom.Sodinokibi - Malwarebytes Labs. `https://blog.malwarebytes.com/detections/ransom-sodinokibi/`. Accessed on 03-06-2021.

[48] Ransomware criminals are targeting US universities. `https://theconversation.com/ransomware-criminals-are-targeting-us-universities-141932`. Accessed on 03-06-2021.

[49] Simplocker – Android Ransomware Alert! `https://blogs.quickheal.com/simplocker-android-ransomware-alert/`. Accessed on 03-06-2021.

[50] System call analysis of Android malware families. `https://sciresol.s3.us-east-2.amazonaws.com/IJST/Articles/2016/Issue-21/Article62.pdf`. Accessed on 03-06-2021.

[51] Take a deep dive into plugx malware. `https://logrhythm.com/blog/deep-dive-into-plugx-malware/`. Accessed on 03-06-2021.

[52] They come, they hide, and they mess up - Android.Obad and Android.Fakedefender. `https://blogs.quickheal.com/they-come-they-hide-and-they-mess-up-android-obad-and-android-fakedefender/`. Accessed on 03-06-2021.

[53] Timeline of IoT malware. `https://www.stratosphereips.org/blog/2020/4/26/timeline-of-iot-malware-version-1`. Accessed on 03-06-2021.

[54] Triout Android Spyware Framework Makes a Comeback, Abusing App with 50 Million Downloads – Bitdefender Labs. `https://labs.bitdefender.com/2019/02/triout-android-spyware-framework-makes-a-comeback-abusing-app-with-50-million-downloads/`. Accessed on 03-06-2021.

[55] Trojan: Android/BaseBridge.A. `https://www.f-secure.com/v-descs/trojan_android_basebridge.shtml`. Accessed on 03-06-2021.

[56] Trojan: Android/Boxer Description | F-Secure Labs. `https://www.f-secure.com/v-descs/trojan/androidboxer.shtml`. Accessed on 03-06-2021.

[57] Trojan: Android/droiddream.a description | f-secure labs. `https://www.f-secure.com/v-descs/trojan/android/droiddream/a.shtml`. Accessed on 11-04-2022.

[58] Trojan: Android/GinMaster.A. `https://www.f-secure.com/v-descs/trojan_android_ginmaster.shtml`. Accessed on 03-06-2021.

[59] Trojan-fakeav. `https://encyclopedia.kaspersky.com/knowledge/trojan-fakeav/`. Accessed on 03-06-2021.

[60] Virustotal-free online virus, malware and url scanner. `https://www.virustotal.com/en`. Accessed on 29-04-2022.

[61] Warning: Crypto-Currency Mining is Targeting Your Android. `https://www.mcafee.com/blogs/mobile-security/warning-crypto-currency-mining-targeting-android/`. Accessed on 03-06-2021.

[62] A whale of a tale: Hummingbad returns - check point software. `https://blog.checkpoint.com/2017/01/23/hummingbad-returns/`. Accessed on 03-06-2021.

[63] What are the different protection levels in Android permissions? `https://blog.mindorks.com/what-are-the-different-protection-levels-in-android-permission`. Accessed on 03-06-2021.

[64] What is a botnet? | kaspersky. `https://www.kaspersky.co.in/resource-center/threats/botnet-attacks`. Accessed on 03-06-2021.

[65] What is an API call? `https://rapidapi.com/blog/api-glossary/api-call/`. Accessed on 03-06-2021.

[66] What is the Joker malware? Here's how it affects apps - The Week. `https://www.theweek.in/news/sci-tech/2020/07/13/What-is-the-Joker-malware-Heres-how-it-affects-apps.html`. Accessed on 03-06-2021.

[67] Zitmo (Zeus-in-the-mobile) Trojan Attacks Android and Blackberry Smartphones. `https://www.enigmasoftware.com/zitmo-zeus-in-the-mobile-trojan-attacks-android-blackberry-smartphones`. Accessed on 11-04-2021.

[68] Yousra Aafer, Wenliang Du, and Heng Yin. Droidapiminer: Mining API-level features for robust malware detection in android. In *Security and Privacy in Communication Networks*, pages 86–103. Springer, 2013.

[69] Moutaz Alazab, Mamoun Alazab, Andrii Shalaginov, Abdelwadood Mesleh, and Albara Awajan. Intelligent mobile malware detection using permission requests and API calls. *Future Generation Computer Systems*, 107:509–521, 2020.

[70] Aylin Alin. Multicollinearity. *Wiley Interdisciplinary Reviews: Computational Statistics*, 2(3):370–374, 2010.

[71] Abdelfattah Amamra, Jean-Marc Robert, Andrien Abraham, and Chamseddine Talhi. Generative versus discriminative classifiers for android anomaly-based detection system using system calls filtering and abstraction process. *Security and Communication Networks*, 9(16):3483–3495, 2016.

[72] Abdelfattah Amamra, Jean-Marc Robert, and Chamseddine Talhi. Enhancing malware detection for android systems using a system call filtering and abstraction process. *Security and Communication Networks*, 8(7):1179–1192, 2015.

[73] Anshul Arora, Shree Garg, and Sateesh K Peddoju. Malware detection using network traffic analysis in Android based mobile devices. In *2014 Eighth International Conference on Next Generation Mobile Apps, Services and Technologies*, pages 66–71. IEEE, 2014.

[74] Anshul Arora and Sateesh K Peddoju. Ntpdroid: A hybrid Android malware detector using network traffic and system permissions. In *2018 17th IEEE International Conference On Trust, Security And Privacy In Computing And Communications)*, pages 808–813. IEEE, 2018.

[75] Daniel Arp, Michael Spreitzenbarth, Malte Hubner, Hugo Gascon, Konrad Rieck, and CERT Siemens. Drebin: Effective and explainable detection of Android malware in your pocket. In *Proceedings of NDSS symposium*, pages 23–26, 2014.

[76] Saba Arshad, Munam A Shah, Abdul Wahid, Amjad Mehmood, Houbing Song, and Hongnian Yu. SAMADroid: A Novel 3-Level Hybrid Malware Detection Model for Android Operating System. *IEEE Access*, 6:4321–4339, 2018.

[77] Nitay Artenstein and Idan Revivo. Man in the binder: He who controls ipc, controls the droid. In *Europe Blackhat Conference*, 2014.

[78] Alessandro Bacci, Alberto Bartoli, Fabio Martinelli, Eric Medvet, Francesco Mercaldo, and Corrado Aaron Visaggio. Impact of code obfuscation on android malware detection based on static and dynamic analysis. In *ICISSP*, pages 379–385, 2018.

[79] Hongpeng Bai, Nannan Xie, Xiaoqiang Di, and Qing Ye. FAMD: A fast multifeature Android malware detection framework, design, and implementation. *IEEE Access*, 8:194729–194740, 2020.

[80] Nazanin Bakhshinejad and Ali Hamzeh. A new compression based method for Android malware detection using opcodes. *19th CSI International Symposium on Artificial Intelligence and Signal Processing, AISP 2017*, pages 256–261, 2018.

[81] Adam Bannister. Intent redirection vulnerabilities in popular android apps spotlight danger of dynamic code loading, warn researchers. https://portswigger.net/daily-swig/intent-redirection-vulnerabilities-in-popular-android-apps-spotlight-danger-of-dynamic-code-loading-warn-researchers. Accessed on 03-06-2021.

[82] Lingfeng Bao, Tien-Duy B Le, and David Lo. Mining sandboxes: Are we there yet? In *2018 IEEE 25th International Conference on Software Analysis, Evolution and Reengineering (SANER)*, pages 445–455. IEEE, 2018.

[83] Mario Luca Bernardi, Marta Cimitile, Damiano Distante, Fabio Martinelli, and Francesco Mercaldo. Dynamic malware detection and phylogeny analysis using process mining. *International Journal of Information Security*, 18(3):257–284, 2019.

[84] Massimo Bernaschi, Emanuele Gabrielli, and Luigi V Mancini. Operating system enhancements to prevent the misuse of system calls. In *Proceedings of the 7th ACM Conference on Computer and Communications Security*, pages 174–183, 2000.

[85] Shweta Bhandari, Rekha Panihar, Smita Naval, Vijay Laxmi, Akka Zemmari, and Manoj Singh Gaur. SWORD: semantic aware Android malware detector. *Journal of Information Security and Applications*, 42:46–56, 2018.

[86] Parnika Bhat and Kamlesh Dutta. A survey on various threats and current state of security in Android platform. *ACM Computing Surveys (CSUR)*, 52(1):1–35, 2019.

[87] Boris Brizzi. Graph-classification-with-gcn. `https://github.com/BrizziB/Graph-Classification-with-GCN`. Accessed on 03-06-2021.

[88] Michael M. Bronstein, Joan Bruna, Yann LeCun, Arthur Szlam, and Pierre Vandergheynst. Geometric deep learning: Going beyond euclidean data. *IEEE Signal Processing Magazine*, 34(4):18–42, 2017.

[89] Iker Burguera, Urko Zurutuza, and Simin Nadjm-Tehrani. Crowdroid: behavior-based malware detection system for android. In *Proceedings of the 1st ACM Workshop on Security and Privacy in Smartphones and Mobile Devices*, pages 15–26, 2011.

[90] Davide Canali, Andrea Lanzi, Davide Balzarotti, Christopher Kruegel, Mihai Christodorescu, and Engin Kirda. A quantitative study of accuracy in system call-based malware detection. In *Proceedings of the 2012 International Symposium on Software Testing and Analysis*, pages 122–132, 2012.

[91] Gerardo Canfora, Eric Medvet, Francesco Mercaldo, and Corrado Aaron Visaggio. Detecting Android malware using sequences of system calls. In *Proceedings of the 3rd International Workshop on Software Development Lifecycle for Mobile*, pages 13–20, 2015.

[92] Victor Chebyshev. Mobile malware evolution 2019. `https://securelist.com/mobile-malware-evolution-2019/`. Accessed on 03-06-2021.

[93] Sen Chen, Minhui Xue, Zhushou Tang, Lihua Xu, and Haojin Zhu. Stormdroid: A streaminglized machine learning-based system for detecting Android malware. In *Proceedings of the 11th ACM on Asia Conference on Computer and Communications Security*, pages 377–388. ACM, 2016.

[94] Tieming Chen, Qingyu Mao, Yimin Yang, Mingqi Lv, and Jianming Zhu. Tinydroid: a lightweight and efficient model for Android malware detection and classification. *Mobile Information Systems*, 2018.

[95] Xiao Chen, Chaoran Li, Derui Wang, Sheng Wen, Jun Zhang, Surya Nepal, Yang Xiang, and Kui Ren. Android HIV: A Study of Repackaging Malware for Evading Machine-Learning Detection. *IEEE Transactions on Information Forensics and Security*, 15:987–1001, August 2018.

[96] Elliot J. Chikofsky and James H Cross. Reverse engineering and design recovery: A taxonomy. *IEEE software*, 7(1):13–17, 1990.

[97] C Chow and Cong Liu. Approximating discrete probability distributions with dependence trees. *IEEE Transactions on Information Theory*, 14(3):462–467, 1968.

[98] code.google.com. Androguard. `http://code.google.com/p/androguard/`. Accessed on 03-06-2021.

[99] Santanu Kumar Dash, Guillermo Suarez-Tangil, Salahuddin Khan, Kimberly Tam, Mansour Ahmadi, Johannes Kinder, and Lorenzo Cavallaro. Droidscribe: Classifying Android malware based on runtime behavior. In *2016 IEEE Security and Privacy Workshops (SPW)*, pages 252–261. IEEE, 2016.

[100] Marko Dimjašević, Simone Atzeni, Ivo Ugrina, and Zvonimir Rakamaric. Evaluation of Android malware detection based on system calls. In *Proceedings of the 2016 ACM on International Workshop on Security And Privacy Analytics*, IWSPA '16, pages 1–8, New York, USA, 2016. ACM.

[101] Karim O Elish, Xiaokui Shu, Danfeng Daphne Yao, Barbara G Ryder, and Xuxian Jiang. Profiling user-trigger dependence for Android malware detection. *Computers & Security*, 49:255–273, 2015.

[102] Ming Fan, Jun Liu, Xiapu Luo, Kai Chen, Zhenzhou Tian, Qinghua Zheng, and Ting Liu. Android malware familial classification and representative sample selection via frequent subgraph analysis. *IEEE Transactions on Information Forensics and Security*, 13(8):1890–1905, 2018.

[103] Zheran Fang, Weili Han, and Yingjiu Li. Permission based android security: Issues and countermeasures. *Computers & Security*, 43:205–218, 2014.

[104] Parvez Faruki, Ammar Bharmal, Vijay Laxmi, Vijay Ganmoor, Manoj Singh Gaur, Mauro Conti, and Muttukrishnan Rajarajan. Android security: a survey of issues, malware penetration, and defenses. *IEEE Communications Surveys & Tutorials*, 17(2):998–1022, 2014.

[105] Ali Feizollah, Nor Badrul Anuar, Rosli Salleh, Guillermo Suarez-Tangil, and Steven Furnell. Androdialysis: Analysis of Android intent effectiveness in malware detection. *Computers & Security*, 65:121–134, 2017.

[106] Adrienne Porter Felt, Erika Chin, Steve Hanna, Dawn Song, and David Wagner. Android permissions demystified. In *Proceedings of the 18th ACM Conference on Computer and Communications Security*, pages 627–638. ACM, 2011.

[107] Pengbin Feng, Jianfeng Ma, Cong Sun, Xinpeng Xu, and Yuwan Ma. A novel dynamic Android malware detection system with ensemble learning. *IEEE Access*, 6:30996–31011, 2018.

[108] André Prata Ferreira, Chetna Gupta, Pedro R. M. Inácio, and Mário M. Freire. Behaviour-based malware detection in mobile android platforms using machine learning algorithms. *Journal of Wireless Mobile Networks, Ubiquitous Computing, and Dependable Applications (JoWUA)*, 12(4):62–88, 2021.

[109] Stephanie Forrest, Steven A Hofmeyr, Anil Somayaji, and Thomas A Longstaff. A sense of self for unix processes. In *Proceedings 1996 IEEE Symposium on Security and Privacy*, pages 120–128. IEEE, 1996.

[110] Jerome H Friedman.   On bias, variance, 0/1—loss, and the curse-of-dimensionality. *Data Mining and Knowledge Discovery*, 1(1):55–77, 1997.

[111] Nir Friedman, Dan Geiger, and Moises Goldszmidt. Bayesian network classifiers. *Machine learning*, 29(2):131–163, 1997.

[112] Laura Gheorghe, Bogdan Marin, Gary Gibson, Lucian Mogosanu, Razvan Deaconescu, Valentin-Gabriel Voiculescu, and Mihai Carabas.   Smart malware detection on android.   *Security and Communication Networks*, 8(18):4254–4272, 2015.

[113] github.com.   Collection of Android malware samples. `https://github.com/killvxk/Android-Malwares-1`. Accessed on 03-04-2021.

[114] github.com.   Collection of Android malware samples. `https://github.com/ashishb/android-malware`. Accessed on 03-06-2021.

[115] github.com. Ransomware. `https://github.com/Sh1n0g1/Ransomware`. Accessed on 18-01-2018.

[116] Aric Hagberg, Pieter Swart, and Daniel S Chult.   Exploring network structure, dynamics, and function using networkx. Technical report, Los Alamos National Lab.(LANL), Los Alamos, NM (United States), 2008.

[117] Hyoil Han, SeungJin Lim, Kyoungwon Suh, Seonghyun Park, Seong-je Cho, and Minkyu Park. Enhanced Android malware detection: An svm-based machine learning approach. In *2020 IEEE International Conference on Big Data and Smart Computing (BigComp)*, pages 75–81. IEEE, 2020.

[118] Juan Manuel Harán. Malware of the 1980s: Looking back at the brain virus and the morris worm. `https://www.welivesecurity.com/2018/11/05/malware-1980s-brain-virus-morris-worm/`. Accessed on 04-11-2021.

[119] Gaofeng He, Bingfeng Xu, and Haiting Zhu. Appfa: a novel approach to detect malicious android applications on the network. *Security and Communication Networks*, 2018.

[120] Robert Hecht-Nielsen. Theory of the backpropagation neural network. In *Neural Networks for Perception*, pages 65–93. 1992.

[121] Seyyedali Hosseinalipour, Jie Wang, Yuanzhe Tian, and Huaiyu Dai. Infection analysis on irregular networks through graph signal processing. *IEEE Transactions on Network Science and Engineering*, 2019.

[122] Shifu Hou, Aaron Saas, Lifei Chen, and Yanfang Ye. Deep4maldroid: A deep learning framework for Android malware detection based on linux kernel system call graphs. In *2016 IEEE/WIC/ACM International Conference on Web Intelligence Workshops (WIW)*, pages 104–111. IEEE, 2016.

[123] Shifu Hou, Aaron Saas, Yanfang Ye, and Lifei Chen. Droiddelver: An Android malware detection system using deep belief network based on api call blocks. In *International Conference on Web-Age Information Management*, pages 54–66. Springer, 2016.

[124] Syed Jawad Hussain, Usman Ahmed, Humera Liaquat, Shiba Mir, NZ Jhanjhi, and Mamoona Humayun. Imiad: intelligent malware identification for Android platform. In *2019 International Conference on Computer and Information Sciences (ICCIS)*, pages 1–6. IEEE, 2019.

[125] Quentin Jerome, Kevin Allix, Radu State, and Thomas Engel. Using opcode-sequences to detect malicious android applications. In *2014 IEEE International Conference on Communications (ICC)*, pages 914–919. IEEE, 2014.

[126] Teenu S John and Tony Thomas. Adversarial attacks and defenses in malware detection classifiers. In *Handbook of Research on Cloud Computing and Big Data Applications in IoT*, pages 127–150. 2019.

[127] Teenu S. John and Tony Thomas. Evading Static and Dynamic Android malware Detection Mechanisms. *Communications in Computer and Information Science*, 1364:33–48, 2021.

[128] Teenu S John, Tony Thomas, and Sabu Emmanuel. Graph convolutional networks for Android malware detection with system call graphs. In *2020 Third ISEA Conference on Security and Privacy (ISEA-ISAP)*, pages 162–170. IEEE, 2020.

[129] Kartik Khariwal, Jatin Singh, and Anshul Arora. Ipdroid: Android malware detection using intents and permissions. In *2020 Fourth World Conference on Smart Trends in Systems, Security and Sustainability (WorldS4)*, pages 197–202. IEEE, 2020.

[130] TaeGuen Kim, BooJoong Kang, and Eul Gyu Im. Runtime detection framework for android malware. *Mobile Information Systems*, 2018: 1–15, HIndawi.

[131] Taeguen Kim, Boojoong Kang, Mina Rho, Sakir Sezer, and Eul Gyu Im. A multimodal deep learning method for Android malware detection using various features. *IEEE Transactions on Information Forensics and Security*, 14(3):773–788, 2019.

[132] Diederik P Kingma and Jimmy Ba. Adam: A method for stochastic optimization. *arXiv preprint arXiv:1412.6980*, 2014.

[133] Thomas N. Kipf and Max Welling. Semi-supervised classification with graph convolutional networks. *5th International Conference on Learning Representations, ICLR 2017 - Conference Track Proceedings*, 2019.

[134] Clemens Kolbitsch, Paolo Milani Comparetti, Christopher Kruegel, Engin Kirda, Xiao-yong Zhou, and XiaoFeng Wang. Effective and efficient malware detection at the end host. In *USENIX Security Symposium*, volume 4, pages 351–366, 2009.

[135] KPMG. The rise of ransomware during covid -19. `https://home.kpmg/xx/en/home/insights/2020/05/rise-of-ransomware-during-covid-19.html`. Accessed on 03-06-2021.

[136] Iggy Krajci and Darren Cummings. History and evolution of the Android OS. In *Android on x86*, pages 1–8. Springer, 2013.

[137] A Lakshmanarao and M Shashi. Android malware detection with deep learning using rnn from opcode sequences. *International Journal of Interactive Mobile Technologies*, 16(1), 2022.

[138] Saskia Le Cessie and Johannes C Van Houwelingen. Ridge estimators in logistic regression. *Journal of the Royal Statistical Society: Series C (Applied Statistics)*, 41(1):191–201, 1992.

[139] Deqiang Li and Qianmu Li. Adversarial deep ensemble: Evasion attacks and defenses for malware detection. *IEEE Transactions on Information Forensics and Security*, 15:3886–3900, 2020.

[140] Shanhong Liu. Android operating system share worldwide by os version from 2013 to 2020. `https://www.statista.com/statistics/271774/share-of-android-platforms-on-mobile-devices-with-android-os/`. Accessed on 03-06-2021.

[141] Federico Maggi, Matteo Matteucci, and Stefano Zanero. Detecting intrusions through system call sequence and argument analysis. *IEEE Transactions on Dependable and Secure Computing*, 7(4):381–395, 2008.

[142] Robert Layton Manoun Alazab, Sitalakshmi Venkataraman, Paul Watters, Mamoun Alazab, and Robert Layton. Malware detection based on structural and behavioural features of api calls. `http://citeseerx.ist.psu.edu/viewdoc/summary?doi=10.1.1.470.6890`, 2010. Accessed on 03-06-2021.

[143] Alejandro Martín, Raúl Lara-Cabrera, and David Camacho. Android malware detection through hybrid features fusion and ensemble classifiers: The andropytool framework and the omnidroid dataset. *Information Fusion*, 52:128–142, 2019.

[144] McAfee. What is fileless malware? https://www.mcafee.com/enterprise/en-in/security-awareness/ransomware/what-is-fileless-malware.html. Accessed on 03-06-2021.

[145] Stuart McIlroy, Nasir Ali, and Ahmed E Hassan. Fresh apps: an empirical study of frequently-updated mobile apps in the google play store. *Empirical Software Engineering*, 21(3):1346–1370, 2016.

[146] Niall McLaughlin. Malceiver: Perceiver with hierarchical and multi-modal features for Android malware detection. *arXiv preprint arXiv:2204.05994*, 2022.

[147] Marisa Mdler. Ransomware as a service (raas) threats. https://insights.sei.cmu.edu/sei_blog/2020/10/ransomware-as-a-service-raas-threats.html. Accessed on 03-06-2021.

[148] Reto Meier. *Professional Android for Application Development*. John Wiley & Sons, 2012.

[149] Allie Mellon. Fileless malware 101: Understanding non-malware attacks allie mellen. https://www.cybereason.com/blog/fileless-malware. Accessed on 27-06-2021.

[150] Atif M Memon and Ali Anwar. Colluding apps: Tomorrow's mobile malware threat. *IEEE Security & Privacy*, 13(6):77–81, 2015.

[151] Microsoft. How malware can infect your pc. https://support.microsoft.com/en-us/windows/how-malware-can-infect-your-pc-872bf025-623d-735d-1033-ea4d456fb76b. Accessed on 04-11-2021.

[152] Jelena Milosevic, Miroslaw Malek, and Alberto Ferrante. A friend or a foe? detecting malware using memory and cpu features. In *SECRYPT*, pages 73–84, 2016.

[153] Peter Mooney and Padraig Corcoran. Using osm for lbs–an analysis of changes to attributes of spatial objects. In *Advances in Location-Based Services*, pages 165–179. 2012.

[154] Annamalai Narayanan, Mahinthan Chandramohan, Lihui Chen, and Yang Liu. Context-aware, adaptive, and scalable Android malware detection through online learning. *IEEE Transactions on Emerging Topics in Computational Intelligence*, 1(3):157–175, 2017.

[155] Gonzalo Navarro. A guided tour to approximate string matching. *ACM Computing Surveys*, 33(1):31–88, 2001.

[156] Thanh-Tuan Nguyen, Giang Thi Thu Nguyen, Thin Nguyen, and Duc-Hau Le. Graph convolutional networks for drug response prediction. *IEEE/ACM Transactions on Computational Biology and Bioinformatics*, 19(1):146–154, 2022.

[157] Juhani Nieminen. On the centrality in a graph. *Scandinavian Journal of Psychology*, 15(1):332–336, 1974.

[158] Robin Nix and Jian Zhang. Classification of android apps and malware using deep neural networks. In *2017 International Joint Conference on Neural Networks (IJCNN)*, pages 1871–1878. IEEE, 2017.

[159] Jon Oberheide and Charlie Miller. Dissecting the android bouncer. *Summer Conference 2012, New York*, 95:110, 2012.

[160] Lucky Onwuzurike, Mario Almeida, Enrico Mariconti, Jeremy Blackburn, Gianluca Stringhini, and Emiliano De Cristofaro. A family of droids-Android malware detection via behavioral modeling: Static vs dynamic analysis. In *2018 16th Annual Conference on Privacy, Security and Trust (PST)*, pages 1–10. IEEE, 2018.

[161] Antonio Ortega, Pascal Frossard, Jelena Kovačević, José MF Moura, and Pierre Vandergheynst. Graph signal processing: Overview, challenges, and applications. *Proceedings of the IEEE*, 106(5):808–828, 2018.

[162] Jasseca Ortega. Fake joomla! plugin keyscaptcha still in the wild. `https://www.sitelock.com/blog/tag/malicious-plugin/`. Accessed on 03-04-2021.

[163] Keyur K Patel, Sunil M Patel, et al. Internet of things-iot: definition, characteristics, architecture, enabling technologies, application & future challenges. *International Journal of Engineering Science and Computing*, 6(5), 2016: 6122–6131.

[164] Abdurrahman Pektaş and Tankut Acarman. Learning to detect Android malware via opcode sequences. *Neurocomputing*, 396:599–608, 2020.

[165] Feargus Pendlebury, Fabio Pierazzi, Roberto Jordaney, Johannes Kinder, and Lorenzo Cavallaro. {TESSERACT}: Eliminating experimental bias in malware classification across space and time. In *28th USENIX Security Symposium (USENIX Security 19)*, pages 729–746, 2019.

[166] Thanasis Petsas, Giannis Voyatzis, Elias Athanasopoulos, Michalis Polychronakis, and Sotiris Ioannidis. Rage against the virtual machine: hindering dynamic analysis of Android malware. In *Proceedings of the Seventh European Workshop on System Security*, pages 1–6, 2014.

[167] Sebastian Poeplau, Yanick Fratantonio, Antonio Bianchi, Christopher Kruegel, and Giovanni Vigna. Execute this! analyzing unsafe and malicious dynamic code loading in android applications. In *NDSS*, volume 14, pages 23–26, 2014.

[168] Boris Procházka, Tomas Vojnar, and Martin Drahansky. Hijacking the linux kernel. In *Sixth Doctoral Workshop on Mathematical and Engineering Methods in Computer Science (MEMICS'10)–Selected Papers*, 2011.

[169] Ford Quin. New tekya ad fraud found on google play. `https://www.trendmicro.com/en_in/research/20/f/new-tekya-ad-fraud-found-on-google-play.html`. Accessed on 03-06-2021.

[170] JR Raphael. Android versions: A living history from 1.0 to 11. `https://www.computerworld.com/article/3235946/android-versions-a-living-history-from-1-0-to-today.html`. Accessed on 29-04-2022.

[171] Vaibhav Rastogi, Yan Chen, and Xuxian Jiang. Catch me if you can: Evaluating android anti-malware against transformation attacks. *IEEE Transactions on Information Forensics and Security*, 9(1):99–108, 2013.

[172] Payam Refaeilzadeh, Lei Tang, and Huan Liu. Cross-validation. In Ling Liu and M. Tamer Özsu, editors, *Encyclopedia of Database Systems*, pages 1–7, New York, 2016. Springer.

[173] Rick Rogers, John Lombardo, Zigurd Mednieks, and Blake Meike. *Android application development: Programming with the Google SDK*. O'Reilly Media, Inc., 2009.

[174] Peter Jay Salzman, Michael Burian, and Ori Pomerantz. The linux kernel module programming guide. `https://tldp.org/LDP/lkmpg/2.6/html/lkmpg.html`. Accessed on 03-06-21.

[175] Aliaksei Sandryhaila and José MF Moura. Discrete signal processing on graphs. *IEEE Transactions on Signal Processing*, 61(7):1644–1656, 2013.

[176] Riccardo Sartea, Alessandro Farinelli, and Matteo Murari. Secur-ama: active malware analysis based on monte carlo tree search for android systems. *Engineering Applications of Artificial Intelligence*, 87:103303, 2020.

[177] Ryo Sato, Daiki Chiba, and Shigeki Goto. Detecting Android malware by analyzing manifest files. *Proceedings of the Asia-Pacific Advanced Network*, 36(23-31):17, 2013.

[178] Asaf Shabtai, Uri Kanonov, Yuval Elovici, Chanan Glezer, and Yael Weiss. Andromaly : a behavioral malware detection framework for android devices. *Journal of Intelligent Information Systems*, 38(1):161–190, 2012.

[179] Hossain Shahriar, Mahbubul Islam, and Victor Clincy. Android malware detection using permission analysis. In *Southeast Conference, 2017*, pages 1–6. IEEE, 2017.

[180] Zhiyong Shan, Iulian Neamtiu, and Raina Samuel. Self-hiding behavior in Android apps: Detection and characterization. In *Proceedings - International Conference on Software Engineering*, pages 728–739. IEEE Computer Society, 2018.

[181] Madhu K Shankarapani, Subbu Ramamoorthy, Ram S Movva, and Srinivas Mukkamala. Malware detection using assembly and api call sequences. *Journal in Computer Virology*, 7(2):107–119, 2011.

[182] Abhijith Shastry, Murat Kantarcioglu, Yan Zhou, and Bhavani Thuraisingham. Randomizing smartphone malware profiles against statistical mining techniques. In *Data and Applications Security and Privacy XXVI*, pages 239–254. Springer, 2012.

[183] Abraham Silberschatz, Peter Baer Galvin, and Greg Gagne. *Operating System Concepts Essentials*. John Wiley & Sons, Inc., 2014.

[184] Cammilie Singleton. Ransomware 2020: Attack trends affecting organizations worldwide. https://securityintelligence.com/posts/ransomware-2020-attack-trends-new-techniques-affecting-organizations-worldwide/. Accessed on 03-06-2021.

[185] Sourceforge.net. Strace Linux system call tracer. https://sourceforge.net/projects/strace. Accessed on 03-06-21.

[186] Orathai Sukwong, Hyong Kim, and James Hoe. Commercial antivirus software effectiveness: an empirical study. *IEEE Computer Architecture Letters*, 44(03):63–70, 2011.

[187] Roopak Surendran and Tony Thomas. Detection of malware applications from centrality measures of syscall graph. *Concurrency and Computation: Practice and Experience*, 34(10), 2022.

[188] Roopak Surendran, Tony Thomas, and Sabu Emmanuel. Android malware detection mechanism based on bayesian model averaging. In *Proceedings of the 5th International Conference on Advanced Computing, Networking, and Informatics*. Springer, 2017.

[189] Roopak Surendran, Tony Thomas, and Sabu Emmanuel. Gsdroid: Graph signal based compact feature representation for Android malware detection. *Expert Systems with Applications*, 159:113581, 2020.

[190] Roopak Surendran, Tony Thomas, and Sabu Emmanuel. On existence of common malicious system call codes in Android malware families. *IEEE Transactions on Reliability*, 70(1):248–260, 2020.

[191] Roopak Surendran, Tony Thomas, and Sabu Emmanuel. A tan based hybrid model for Android malware detection. *Journal of Information Security and Applications*, 54:102483, 2020.

[192] Richard S Sutton and Andrew G Barto. *Reinforcement learning: An introduction*. MIT press, 2018.

[193] Kabakus Abdullah Talha, Dogru Ibrahim Alper, and Cetin Aydin. Apk auditor: Permission-based android malware detection system. *Digital Investigation*, 13:1–14, 2015.

[194] Kimberly Tam, Salahuddin J Khan, Aristide Fattori, and Lorenzo Cavallaro. Copperdroid: Automatic reconstruction of Android malware behaviors. In *NDSS*, pages 1–15, 2015.

[195] Junwei Tang, Ruixuan Li, Yu Jiang, Xiwu Gu, and Yuhua Li. Android malware obfuscation variants detection method based on multi-granularity opcode features. *Future Generation Computer Systems*, 129:141–151, 2022.

[196] Mobile Threat Response Team. Slocker mobile ransomware starts mimicking wannacry - trendlabs security intelligence blog. https://www.trendmicro.com/en_in/research/17/g/slocker-mobile-ransomware-starts-mimicking-wannacry.html. Accessed on 03-06-2021.

[197] Fei Tong and Zheng Yan. A hybrid approach of mobile malware detection in Android. *Journal of Parallel and Distributed Computing*, 103:22–31, 2017.

[198] Nicolas Viennot, Edward Garcia, and Jason Nieh. A measurement study of google play. In *2014 ACM International Conference on Measurement and Modeling of Computer Systems*, pages 221–233, 2014.

[199] P Vinod, Akka Zemmari, and Mauro Conti. A machine learning based approach to detect malicious android apps using discriminant system calls. *Future Generation Computer Systems*, 94:333–350, 2019.

[200] David Wagner and Paolo Soto. Mimicry attacks on host-based intrusion detection systems. In *Proceedings of the 9th ACM Conference on Computer and Communications Security*, pages 255–264, 2002.

[201] Kelle Wanderlee. They come in the night: Ransomware deployment trends. https://www.fireeye.com/blog/threat-research/2020/03/they-come-in-the-night-ransomware-deployment-trends.html. Accessed on 03-06-2021.

[202] Xinran Wang, Yoon-Chan Jhi, Sencun Zhu, and Peng Liu. Detecting software theft via system call based birthmarks. In *2009 Annual Computer Security Applications Conference*, pages 149–158. IEEE, 2009.

[203] Fengguo Wei, Yuping Li, Sankardas Roy, Xinming Ou, and Wu Zhou. Deep ground truth analysis of current Android malware. In *International Conference on Detection of Intrusions and Malware, and Vulnerability Assessment*, pages 252–276. Springer, 2017.

[204] Lukas Weichselbaum, Matthias Neugschwandtner, Martina Lindorfer, Yanick Fratantonio, Victor van der Veen, and Christian Platzer. Andrubis: Android malware under the magnifying glass. *Vienna University of Technology, Tech. Rep. TRISECLAB-0414-001*, 2014.

[205] Ryan Welton and Grassi Marco. Current state of android privilege escalation. `https://filepursuit.unblockproject.org/file/18925610-D2-Ryan-Welton-and-Marco-Grassi-Current-State-of-Android-Privilege-Escalation-pdf/`. Accessed on 27-04-2022.

[206] Le Wu, Peijie Sun, Richang Hong, Yanjie Fu, Xiting Wang, and Meng Wang. Socialgcn: An efficient graph convolutional network based model for social recommendation. *arXiv preprint arXiv:1811.02815*, 2018.

[207] Zonghan Wu, Shirui Pan, Fengwen Chen, Guodong Long, Chengqi Zhang, and S Yu Philip. A comprehensive survey on graph neural networks. *IEEE Transactions on Neural Networks and Learning Systems*, 32(1):4–24, 2020.

[208] Xi Xiao, Zhenlong Wang, Qing Li, Shutao Xia, and Yong Jiang. Back-propagation neural network on markov chains from system call sequences: a new approach for detecting android malware with system call sequences. *IET Information Security*, 11(1):8–15, 2017.

[209] Xi Xiao, Xianni Xiao, Yong Jiang, Xuejiao Liu, and Runguo Ye. Identifying android malware with system call co-occurrence matrices. *Transactions on Emerging Telecommunications Technologies*, 27(5):675–684, 2016.

[210] Xi Xiao, Shaofeng Zhang, Francesco Mercaldo, Guangwu Hu, and Arun Kumar Sangaiah. Android malware detection based on system call sequences and lstm. *Multimedia Tools and Applications*, 78(4):3979–3999, 2019.

[211] Liang Xie, Xinwen Zhang, Jean-Pierre Seifert, and Sencun Zhu. PBMDS: a behavior-based malware detection system for cellphone devices. In *Proceedings of the Third ACM Conference on Wireless Network Security*, pages 37–48. ACM, 2010.

[212] Ke Xu, Yingjiu Li, Robert Deng, Kai Chen, and Jiayun Xu. Droidevolver: Self-evolving Android malware detection system. In *2019 IEEE European Symposium on Security and Privacy (EuroS&P)*, pages 47–62. IEEE, 2019.

[213] Ke Xu, Yingjiu Li, and Robert H Deng. Iccdetector: Icc-based malware detection on android. *IEEE Transactions on Information Forensics and Security*, 11(6):1252–1264, 2016.

[214] Lifan Xu, Dongping Zhang, Nuwan Jayasena, and John Cavazos. Hadm: Hybrid analysis for detection of malware. In *Proceedings of SAI Intelligent Systems Conference*, pages 702–724. Springer, 2016.

[215] Xin Xu. Sequential anomaly detection based on temporal-difference learning: Principles, models and case studies. *Applied Soft Computing*, 10(3):859–867, 2010.

[216] Wei Yu, Hanlin Zhang, Linqiang Ge, and Rommie Hardy. On behavior-based detection of malware on android platform. In *2013 IEEE Global Communications Conference (GLOBECOM)*, pages 814–819. IEEE, 2013.

[217] Zhenlong Yuan, Yongqiang Lu, and Yibo Xue. Droiddetector: Android malware characterization and detection using deep learning. *Tsinghua Science and Technology*, 21(1):114–123, 2016.

[218] Mu Zhang, Yue Duan, Heng Yin, and Zhiruo Zhao. Semantics-aware Android malware classification using weighted contextual api dependency graphs. In *Proceedings of the 2014 ACM SIGSAC Conference on Computer and Communications Security*, pages 1105–1116, 2014.

[219] Shaofeng Zhang and Xi Xiao. Cscdroid: Accurately detect Android malware via contribution-level-based system call categorization. In *2017 IEEE Trustcom/BigDataSE/ICESS*, pages 193–200. IEEE, 2017.

[220] Ling Zhao, Yujiao Song, Chao Zhang, Yu Liu, Pu Wang, Tao Lin, Min Deng, and Haifeng Li. T-gcn: A temporal graph convolutional network for traffic prediction. *IEEE Transactions on Intelligent Transportation Systems*, 21(9):3848–3858, 2019.

[221] Min Zheng, Mingshen Sun, and John CS Lui. Droidtrace: A ptrace based android dynamic analysis system with forward execution capability. In *2014 International Wireless Communications and Mobile Computing Conference (IWCMC)*, pages 128–133. IEEE, 2014.

[222] Yajin Zhou and Xuxian Jiang. Dissecting android malware: Characterization and evolution. In *2012 IEEE Symposium on Security and Privacy*, pages 95–109. IEEE, 2012.

Index